FA-159, 2/96

# TECHNICAL RESCUE PROGRAM DEVELOPMENT MANUAL

United States
Fire Administration

Federal Emergency
Management Agency

## August 1995

This publication was produced under contract EMW-94-C 4381 for the U.S. Fire Administration, Federal Emergency Management Agency. Any information, findings, conclusions, or recommendations expressed in this publication do not necessarily reflect the views of the Federal Emergency Management Agency or the U.S. Fire Administration.

# ACKNOWLEDGMENTS

A large portion of the information contained in this manual comes directly from fire service personnel who have been involved in technical rescue teams themselves. The principle contributors were: Chief Michael Tamillow, Fairfax County (VA) Fire and Rescue Department; Sergeant John Bentivoglio, Bethesda-Chevy Chase (MD) Rescue Squad; Chief Michael Iacona, West Palm Beach (FL) Fire Department; Kim Houser, Westmoreland (PA) Department of Emergency Management; Chief Pete West, Fair Oaks (VA) Volunteer Fire and Rescue Department; and Captain Tim Gallagher, Phoenix (AZ) Fire Department. Individuals who assisted in the review of the manual include: Lieutenant Giovannie Ulloa, Metro-Dade County (FL) Fire Department; Chief John Eversole, Chicago (IL) Fire Department; Mark Ghalarducci, California Office of Emergency Services; Lieutenant Steve Stein, Butler Township (OH) Fire Department; and Chief Steve Lancaster, Baltimore County (MD) Fire Department.

The U.S. Fire Administration would like to thank the many fire department personnel and rescue organizations who kindly shared their experience in and knowledge about technical rescue for inclusion in this report.

# TABLE OF CONTENTS

# INTRODUCTION

F ire departments across the United States have assumed a major role as primary responders to rescue incidents that involve, among other things, structural collapse, trench cave-in, confined spaces, industrial and agricultural machinery water emergencies, and people trapped above or below grade level. These emergencies are grouped into a category of rescue called technical rescue. Technical rescue incidents are often complex, requiring specially trained personnel and special equipment to complete the mission. Natural forces such as earth tremors, precipitation, temperature extremes, and swift water currents often complicate technical rescue incidents. The presence of flammable vapors and toxic chemicals can also increase the level of risk. The safety of crews conducting technical rescue operations is of special concern.

Fire departments and rescue squads throughout the country perform technical rescues on a daily basis. Some complex technical rescue incidents last many hours or even days as rescue personnel carefully assess the situation, obtain and set up the appropriate rescue equipment, monitor scene safety, and remove hazards before they can finally reach, stabilize, and extricate the victims. The presence of hazards such as flammable vapors or dust often forces rescuers to take additional precautions and time to ensure that operations are conducted safely. Experience has shown that hasty rescue operations can endanger the lives of both rescuers and victims. At the same time, rescuers know that a victim's survival chances are often dependent on quick extrication and transportation to a hospital.

Some departments are better prepared than others to perform technical rescue operations. To deal with these complicated rescue operations, many fire departments have created special technical rescue teams. A technical rescue team is a specialized group of personnel having advanced training and special equipment to safely and efficiently conduct specialized rescue operations. The specialties and capabilities of individual teams vary greatly, depending on their level of training, number of trained personnel, and availability of specialized rescue tools and equipment. For example, some departments have the training and

equipment to perform rescues at collapsed structures by cutting through concrete and removing heavy debris, while other departments are limited to working with picks and shovels to remove debris.

Many departments have *single-discipline* rescue teams such as a dive rescue team. These teams are trained and equipped to handle one type of rescue. Other departments have *multi-discipline* teams that are prepared to perform more than one type of rescue.

The formation of a functional and safe technical rescue team, whether single- or multi-discipline, requires careful planning, a large time commitment from the team members, equipment research and acquisition, risk analysis, training, and funding. This manual provides guidance on how to form a technical rescue team. It discusses many of the considerations that must be made before forming a team such as:

- *Do we need a team for our community?*
- *What type of team does our community need?*
- *How do we conduct a risk assessment to identify rescue hazards?*
- *How do we start a team?*
- What *training is necessary for team members?*
- *What dangers are involved in technical rescue?*
- *How can we fund the team?*
- *What type of personnel will we need on the team?*
- *What laws and standards pertain to rescue?*
- *What equipment will the team need?*

The U. S. Fire Administration initiated a project in 1994 to write a comprehensive program manual on technical rescue. The project was initiated because of the many requests made by fire and rescue personnel to the U.S. Fire Administration for information about technical rescue programs. Many of the rescuers have sought answers to the questions above as well as information about laws and standards affecting rescuers, the Federal Urban Search and Rescue Program, training requirements, equipment needed for a team, and many other topics. This manual was designed to provide answers to topics such as these.

# CHAPTER 1: OVERVIEW OF TECHNICAL RESCUE CAPABILITIES

**A**t some point in time, almost every rescue agency will be presented with a unique or complex rescue situation requiring special skills and equipment to safely resolve. Some agencies are prepared to handle such events, but in many cases, the skills and equipment needed for these events exceed the capabilities of the responding agency.

Many fire departments and rescue agencies across the country have recognized that their baseline skills and their existing equipment are insufficient for rescue incidents that have occurred or may occur in their response areas. Some departments have formed or considered forming technical rescue teams to address these complex and hazardous situations.

Most newly formed teams begin with training members in a single discipline, such as rope rescue or water rescue. Once this team is developed, it may expand into other areas of rescue so that it is a multi-discipline team which can handle several types of advanced rescue. An agency may also choose to establish different teams with individual capabilities.

Various rescue disciplines exist. The rescue disciplines discussed in this manual include:

• *Confined space rescue.* A confined space is an enclosed area with limited entry or egress, which has an internal configuration not designed for human occupancy such that an entrant could become trapped or asphyxiated. It may have inwardly converging walls or a floor which slopes downward and tapers to a smaller cross section. These spaces include sewers, vats, caves, tanks, and other areas. Rescues from such spaces are dangerous, especially if the interior environment is toxic or oxygen deficient. The Occupational Safety and Health Administration (OSHA) terms these dangerous areas "permit-required confined spaces." OSHA estimates that there are over 240,000 permit-required spaces across the United States.

• *Water and ice rescue.* Rescues from lakes, swamps, flooded areas, swift or calm water bodies, and the ocean fit into this category. There are several different specialties within water rescue including swiftwater, calm water, underwater, surf, and ice rescue. Each of these requires special training unto itself.

• *Collapse rescue.* This involves building collapse or other structural collapse such as the collapse of the elevated highway during the 1989 earthquake in Oakland, California. Many collapse rescue teams have been established in earthquake prone areas. They may also be needed in cities that have many older buildings or new construction projects.

• *Trench/cave-in rescue.* Trench or cave-in rescue could occur in almost any jurisdiction across the country. Trenches are often found in areas of new construction where pipes or cables are being buried. The most common trench rescue scenario involves rescuing a construction worker trapped when the trench walls collapse.

• *Rope rescue.* High-angle or low-angle rescues are likely to occur around cliffs, ravines, caves, mountainous areas, highrise buildings, communications towers, water towers, or silos. These rescues may require complex rope and hauling systems to safely secure personnel and extricate victims.

• *Industrial rescue.* Industrial machinery presents many challenges to rescuers. Many industrial rescues involve confined spaces or heavy extrication to free victims trapped by machinery.

• *Agricultural rescue.* Rural fire departments must be prepared to deal with rescues involving individuals trapped under or inside agricultural machinery or silos.

## THE CONCEPT OF A TIERED RESPONSE SYSTEM

This manual will frequently refer to a concept called "tiered response" that is used by many rescue agencies across the country. The concept is to train and equip personnel or units throughout a department to different response levels, or tiers, from a basic rescue level to an advanced rescue capability.

The basic premise of a tiered response system begins with training all personnel to a basic rescue awareness level that will familiarize them with rescue hazards, dangers, and some basic, practical rescue skills. In the event of a complicated rescue, they will request the response of an advanced team and initiate measures within their capabilities until the advanced team arrives. This tiered response system for technical rescue is similar to a tiered EMS response system that uses a basic emergency medical technician to initiate care until a paramedic arrives on the scene.

As an example, Los Angeles County uses this approach for water rescue. All of its engine companies are trained and equipped to handle a basic water rescue incident. For a more complicated situation, engine company personnel are trained and equipped to initiate basic rescue measures until the advanced water rescue team arrives.

There are several advantages to a tiered response system:

• It provides basic rescue training for all personnel.

• All potential rescuers become more aware of the dangers of different situations and recognize situations that are beyond their capabilities.

• A smaller number of personnel can develop a high level of expertise in a particular area.

• It fits in well with a regional rescue response system, where the personnel with basic training can handle a basic incident on their own and have the option of calling an advanced regional rescue team for assistance.

•It is less expensive and time consuming than equipping and training all personnel to an advanced rescue level.

Table 1-1 describes a tiered rescue system for rope rescue incidents.

### Table 1-1. Example of Tiered Rope Rescue Response System

| Tier or Response Level | Units | Capabilities |
|---|---|---|
| Level 1 | Engines | Basic rappelling only |
| Level 2 | Ladder Trucks | Rappelling; basic hauling systems; lifting systems using ladders; stokes basket rescues |
| Level 3 | Heavy Squads or Technical Rescue Team | Rappelling; advanced hauling systems: entry Into confined spaces using rope systems; cliff rescue: rope stabilization systems; stokes basket rescues |

In the above example, all engine companies in the department are capable of performing a basic rope rescue. If necessary, the personnel on engine companies could rappel to a patient and begin emergency care until a Level 2 or Level 3 team arrives to remove the patient. In a large department, the Level 2 or Level 3 team may be a resource from within the department. In smaller departments, the Level 2 or Level 3 team may come from another jurisdiction or could be a regional team with members from different jurisdictions. An engine company may be able to handle a simple rope rescue incident on its own but the higher level team is available with the advanced training and additional equipment should it be necessary

A collapse rescue team practices concrete debris removal techniques using a pneumatic breaker. *(Courtesy Mike Tamillow)*

Rescuers wearing special supplied air breathing apparatus enter a confined space at an industrial facility. *(Courtesy Mike Tamillow)*

A rescuer from a confined space rescue team is lowered through a manhole into a sewer. Some confined space entry practices are regulated by the Occupational Safety and Health Administration. *(Courtesy Mike Tamillow)*

An advanced trained water rescue team utilizes a helicopter to retrieve a victim and rescuer from swift currents.

Special rigging is used by a confined space team to lower a rescuer into a timber. Similar tactics would be needed to lower a rescuer into a silo or any other confined space. *(Courtesy Mike Tamillow)*

Rescuers raise a victim using a stokes basket and a mechanical advantage hauling system. *(Courtesy Mike Tamillow)*

A FEMA Urban Search and Rescue Task Force sets up its base camp in Homestead, Florida to assist with search and rescue operations following the destruction caused by Hurricane Andrew. *(Courtesy Mike Tamillow)*

A trench collapse team shores up trench walls using struts before beginning a rescue. (Courtesy *Mike Timillow*)

An unshored trench presents many hazards to victims and rescuers, including secondary collapse.

The Miami Valley (OH) Regional Technical Rescue Team prepares to conduct a trench rescue drill. This team is comprised of members from several area departments.

Collapse rescue team members practice concrete debris stabilization and removal techniques. *(Courtesy Mike Tamillow)*

Los Angeles County utilizes a tiered water rescue response system. Above, an engine company practices the deployment of an inflated section of hose used to catch a victim flowing downstream. This rescue technique can be performed by units with personnel trained only to a basic water rescue level using readily available equipment. Below, the department's advanced water rescue team catches a victim using a water rescue net, which requires advanced rigging techniques. (*Courtesy Larry Collins*)

# CHAPTER 2: CONSIDERATIONS BEFORE FORMING A TEAM

**T**his chapter describes the types of factors that must be evaluated when considering whether or not to form a technical rescue team. For the purposes of this manual, a team will refer to a group of persons who are trained and equipped to perform technical rescues in one or more specialized areas.

Many considerations must be made before starting a rescue team, including whether a team is really needed, whether local officials will support a team, whether firefighters are committed to forming a team, what are the risks associated with a rescue team, and what laws affect the formation of a team.

The following questions should be considered by a fire or rescue chief before starting-up a technical rescue team.

## Is a team needed in our community?

This question can be answered by conducting a risk analysis of your community (described in Chapter 3). The ultimate decision for choosing to develop technical rescue expertise should be based on the needs of your community. You must honestly and accurately assess the risk level in your community and if the risk is real, you should make every effort to secure the necessary resources to perform a rescue safely and efficiently If a need does exist, but this need is being satisfied by an outside response team that is available to respond into your jurisdiction, then developing your own team may be unnecessary.

## What type of team is needed for our community?

Another consideration centers on the type of team that would be needed. Should the team have a single function, or is expertise needed in multiple disciplines? Again, this question can best be answered after conducting a risk assessment (described in Chapter 3).

## Do we have commitment from the firefighters and rescuers for this?

You should thoroughly consider the ability of your emergency response personnel to take on a new challenge. The level of commitment needed to start a technical rescue team is extremely high - it requires dedicated leadership and participation on the part of the entire workforce. Many times we look only at the members who are undergoing the training and forget to evaluate the impact of this training on their co-workers who assume additional firefighting workloads during technical rescue-related absences. From this perspective a total commitment and clear understanding of the impact of this responsibility must be shared throughout the department in pursuing a technical rescue responsibility. Several very dedicated individuals will be necessary to coordinate the team and to monitor team equipment and training.

## How much will it cost to form a team, and is funding is available?

You must thoroughly evaluate both the start up costs and the ongoing operational costs for this type of venture. Start up costs may be very expensive, but depend on the equipment you already possess and the type of team you want to initiate. A majority of start up costs go toward equipment purchases and training. Operational costs may include ongoing training, equipment maintenance, and salaries if paid employees are utilized.

You must consider whether the funding already exists for a new rescue team and how likely it is for you to obtain funding. Funding may come internally from your city or externally from donations by outside organizations.

## Would elected officials and city management support a Technical Rescue Team?

The formation of any rescue team will require support and commitment from officials outside of the fire department. They will have the ultimate say about funding a team. The basic expenses such as purchasing special equipment or funding overtime training can only be met if there is full support from outside officials. Their support is also necessary if you try to share resources with other communities.

In many instances the decision by local authorities to develop an expertise in technical rescue is prompted by an incident of significant magnitude in which the local responders were found to be ill prepared to handle the situation. Emergency managers may feel the need to develop technical rescue skills but, in the absence of a major incident, are unsure of how to justify this type of expenditure. Consider the questions that will be asked by fiscal personnel or elected officials about these expenditures such as, "Why do we need all this expensive equipment?" "How many collapses did we have last year?" or "We've done just fine in the past, why do we need it now?"

As an emergency manager you may be acutely aware of the limitations of your current capabilities and the potential criticism that may result if you are not prepared when a major incident occurs. You should recognize the risks that are involved if you commit emergency workers to a work environment that they are insufficiently trained or ill equipped to handle. Consider whether you can explain these risks to managers and elected officials and what their reactions will be. You must be willing to do some basic background research on risks and needs before trying to "sell" the idea of forming a team to outside officials. Be prepared to provide them hard evidence to gain their support.

## Are other resources available from neighboring communities?

As you assess your technical rescue needs, consider the option of sharing these resources among two or more communities, or setting up a team for a single community. Utilizing a shared or multi-agency response is fiscally responsible and can provide an appropriate level of service. Several factors will come into play in choosing the option best suited for your community. This concept is discussed further in Chapter 10.

## What dangers are posed by forming a team?

Technical rescue, like firefighting, is dangerous. Certainly risks can be limited by providing proper training about safe rescue techniques and by purchasing equipment designed to make rescues safer, but you must consider what dangers will confront rescuers and whether you and the rescuers are willing to face these dangers in a real incident.

According to OSHA, 60 percent of the deaths in confined spaces involve would-be rescuers. Technical rescuers may face many risks including asphyxiation within a confined space, fall injuries from operating on ropes, and drowning while operating in swiftwater conditions.

One of the greatest mistakes made when forming a team is to think that you can create a team without basic training and basic equipment. Some departments have attempted to start a team or perform dangerous rescues without having basic equipment or training. This is extremely risky from the standpoint of both the rescuers and the victims.

## What laws, regulations, and standards affect development of a team?

One of the most complicated and misunderstood areas affecting technical rescue is legal mandates and standards. A host of mandates and standards have been written which affect different types of rescues. Certain aspects of technical rescue are regulated by the Occupational Safety and Health Administration (OSHA) or are discussed in standards from organizations such as the National Fire Protection Association (NFPA). Many of these regulations affect confined space rescue. Compliance with these regulations is recommended for all rescuers for safety purposes, but only certain states mandate compliance.

The various regulations and standards affecting technical rescue are explained in Chapter 6. Before starting a team, a fire or rescue chief must consider what laws will affect a team and the costs of compliance and non-compliance. Failure to comply with a law during a rescue can result in fines or other penalties.

## What training requirements exist?

State and Federal training requirements must be considered when planning for a rescue team. Mandatory training requirements vary from state to state or even among localities. Most technical rescue training mandates are self-determined by a state or locality which may require you to follow a particular training standard established by OSHA, the NFPA, a private training organization, or other entity. The NFPA has issued a standard on training for structural collapse responders. A broader training standard covering all disciplines of technical rescue is expected to be issued in the next few years (refer to Chapter 7 for more information).

# CHAPTER 3: HOW TO FORM A TECHNICAL RESCUE TEAM

**T**he formation and development of a technical rescue team is a considerable undertaking. While the formation of all aspects of a team, both administrative and operational, is quite intensive, the maintenance and recurring training is even more challenging. It can be an expensive undertaking requiring new training and equipment, and most importantly, careful planning.

This chapter recommends steps you should take to form a technical rescue team. The steps are organized into four phases of team development:

| PHASE I: ASSESSMENT OF COMMUNITY RISKS AND RESCUE NEEDS |
| --- |
| 1. Perform a Risk Assessment |
| 2. Analyze Data to Project the Likelihood of a Technical Rescue Emergency |
| 3. Establish a Risk Threshold |
| 4. Determine What Type of Team Is Needed |

| PHASE II: PLANNING |
| --- |
| 1. Establish a Planning Committee to Develop a Plan |
| 2. Determine Current Capabilities |
| 3. Prepare a Concept of Operations |
| 4. Determine Program Management Structure |
| 5. Develop a Staffing Plan |
| 6. Identify Initial Equipment and Vehicle Requirements |
| 7. Identify Training Requirements |
| 8. Consider a Plan for Delivering Recurring Training |
| 9. Estimate Cost of Team and Develop a Budget |
| 10. Obtain Management Support |
| 11, Obtain Political Support |
| 12. Look for Partnerships |

| PHASE III: DEVELOPMENT OF TEAM |
| --- |
| 1. Select the Team Members |
| 2. Train the Team |
| 3. Purchase Equipment and Uniforms |
| 4. Purchase Vehicles |
| 5. Provide Administrative Support |

| PHASE IV: DEVELOPMENT OF STANDARD OPERATING PROCEDURES |
| --- |
| 1. Obtain or Write Administrative and Operational SOPs for the Team |
| 2. Review and Revise SOPs Regularly |

assessment of community risks and rescue needs; planning; development of team; and development of standard operating procedures. Because of the complexity of forming a technical rescue team, each step must be carefully considered so that important issues are not missed.

## PHASE I: ASSESSMENT OF COMMUNITY RISKS AND RESCUE NEEDS

| |
| --- |
| 1. Perform a risk assessment |
| 2. Analyze data to project the likelihood of a technical rescue emergency |
| 3. Establish a risk threshold |
| 4. Determine what type of team is needed |

Many departments begin technical rescue teams after a specific rescue incident has shown a deficiency or inability to safely and effectively handle a rescue. In some cases, a team is developed before a major rescue incident occurs due to the expectation of emergencies created by risks in the community.

In determining whether a team is needed in your community, you must first do some research to evaluate the risks in your area. A risk analysis will help you determine what the level of risk is and what potential hazards exist so that you can decide whether a team is really needed. This is a particularly important part of starting a team for two reasons. First, political leaders will want to know what risks exist to justify funding a team. Second, you will want to know what risks confront your department, what type of hazardous scenarios to train for, and what rescue equipment will be needed to address the risks. A thorough risks analysis should define your objective for a team and justify the effort of forming a team.

### 1. Perform a Risk Assessment

A risk assessment should be based on historical data on rescues plus an analysis of newly introduced risks.

Begin by assessing past rescue needs in your response area. You may look to incident reports to determine the frequency of technical rescues in your area. Other potential sources of data include your state workers' compensation office; your state OSHA or national OSHA; construction or contractors' associations; building officials and inspectors; and safety managers at local businesses. Past experience may indicate the likelihood of technical rescue-type incidents during major construction projects.

You must also consider target hazards that exist in your response area now or you anticipate in the future. Target hazards are specific risk areas that confront your department in a rescue emergency (see Figure 3-1). A review of the natural features of a locality will reveal some hazards. Rivers, rapids, cliffs and rock climbing sites are but a few of the areas where incidents may occur. A review of existing pre-fire plans may highlight certain types of commercial or industrial facilities that might require the services of a specialty team. Contact your local building department to identify new or planned construction which may contain target hazards.

Make a list of target hazards which present special rescue challenges requiring special technical rescue equipment or advanced rescue training to safely and effectively control. Survey department personnel about their knowledge of hazards. You may want to perform inspections of facilities or areas in your jurisdiction which are likely to contain hazards. This may be done in conjunction with your fire safety inspections. You also can mail a written questionnaire to industrial facilities or plant managers asking them what hazards exist at their facilities. Be sure to communicate to them the specific hazards which concern your department.

Regardless of the size or economic make-up of the community almost every jurisdiction is subject to some type risk, such as a major transportation accident or construction collapse, that would necessitate technical rescue expertise. The prevalence or concentration of a specific industry in a community may guide emergency officials to prioritize and develop expertise in areas of technical rescue that have the greatest likelihood for generating an occurrence with that type of industry or activity. Remember that the justification for an expenditure would appear more reasonable and would be more likely to be approved if it can be related to target hazards and risk potentials.

## 2. Analyze Data to Project the Likelihood of a Technical Rescue Emergency

To demonstrate the likelihood of a technical rescue incident, begin by showing the frequency, or rate of which incidents have occurred in a given period of time in your community or even in other

*Figure 3-1. Common Risks and Target Hazards Found in Communities*

| Risk | Potential Hazards Posed by Risk |
|------|--------------------------------|
| Sewers | Confined spaces, toxic gases, oxygen deficiency |
| Rivers/flood ducts, flood-prone areas | Swift water rescue, calm water rescue, toxic water environments, surface and underwater rescue, ice rescue |
| Industrial facilities | Hazardous materials, toxic gas emissions, confined spaces, machinery entrapment |
| Cliffs/gorges/ ravines/mountains | Above grade and below grade rescue |
| Agricultural facilities | Dust explosions, confined spaces, hazards materials, fertilizers, machinery entrapment |
| Cesspools/tanks | TOXIC gases, oxygen deficiency, confined spaces |
| New construction | Structural collapse, trench rescue, machinery entrapment |
| Old buildings | Structural collapse |
| Wells/caves | Confined spaces, hazardous environments |
| Highrises | High angle rescue, elevator rescue |
| Earthquakes/hurricanes/ tornados | Collapse rescue, extrication, disaster response |
| Transfer facilities | Hazardous materials. toxic gas emissions, confined spaces, machinery entrapment |
| Transportation centers | Hazardous materials, toxic gas emissions, confined spaces, machinery entrapment, derailments |

communities with similar hazards as your community has. By demonstrating in simple terms that a reasonable likelihood exists for a particular type of rescue to occur, you will find that elected officials and administrators are more likely to support these enhancements to your readiness.

## EXAMPLE

The town below has had a rapidly growing industrial base since 1980. Between 1980 and 1990, the number of confined space hazards has tripled from 15 to 40 spaces. In the same period, the frequency of confined space rescues has increased.

| Year | Number of Confined Spaces at Indusfries Within the Community | Number of Confined Space Emergency Incidenfs Per Year |
|------|------|------|
| 1980 | 15 | 2 |
| 1981 | 15 | 0 |
| 1982 | 17 | 3 |
| 1983 | 18 | 2 |
| 1984 | 18 | 4 |
| 1985 | 22 | 5 |
| 1986 | 24 | 3 |
| 1987 | 28 | 7 |
| 1988 | 31 | 9 |
| 1989 | 35 | 12 |
| 1990 | 40 | 18 |

In 1980, there were 2 confined space emergencies among the 15 hazard spaces. In 1985, there were 5 emergencies among the 22 hazard spaces, and in 1990, 18 emergencies among the 40 hazard spaces. This data shows that as the number of target hazards have increased, the frequency of confined space emergency incidents per target hazard has also increased.

You can project the likely number of incidents in the future by estimating the number of hazardous spaces in future years, and then multiplying this by the rate of incidents per space. This example of hard data is the evidence that will be necessary to present to managers or the city council to demonstrate the need for a technical rescue team.

## 3. Establish a Risk Threshold

The final determination in a risk assessment should involve weighing the potential risk to the community and the potential risk to emergency responders who must perform the rescues. The presence of hazards in a community creates a risk that someone will become injured or need assistance from rescuers. Likewise, if the community expects the fire department to provide rescue assistance, the lives of firefighters performing a rescue will be put at risk.

Risks vary in severity. The presence of one risk may be very mild, whereas the presence of another very severe. The severity of a hazard must be considered as part of a final risk determination. In terms of a water rescue team, the risks created by a small pond are much less than those created by a swiftwater channel. Likewise, the probability of the occurrence of a rescue incident involving a swiftwater channel is usually greater than that involving a small pond. The community with the small pond may determine that the risk level created by the pond is too minor to warrant a special water rescue team, whereas the community with the swiftwater channel may determine otherwise. If the firefighters or rescuers are expected to perform rescues in hazardous environments, they will face risks including toxic environments and inhalation injuries (confined space rescue), drowning (water rescue), falls (rope rescue), secondary collapses and crush syndrome (collapse rescue), and explosions (silo rescue).

Each community will have to make its own determination about what an acceptable level of risk is, and what is the risk "threshold" that will necessitate the formation of a special rescue team. The community and city managers should know exactly what the fire department's rescue capabilities and limitations are, what risks confront the community, and the dangers that rescuers face in performing rescues. The community should not expect rescuers to perform certain rescues without proper training and equipment.

In most cases, the public will not expect the fire department to rescue everyone in any predicament. Firefighters should not be expected to perform "suicide" rescue missions; however, when there is no forethought or when rescue operations are clearly botched, there is likely to be public outcry.

### 4. Determine What Type of Team Is Needed

The risk analysis should help you determine whether a team is actually necessary. If you decide a team is needed, the next step is to determine what type of team is needed. What risks are you trying to address? Will the team handle only basic rescues or will it be expected to perform complex rescues? What types of emergencies will this team respond to? Define the extent of the capabilities you think you need to provide. These may include: high angle/rope rescue, trench collapse, structural collapse, confined space, agricultural rescue, machinery entrapment, calm or swift water rescue, or dive rescue.

Remember that if a multi-discipline team is needed to cover several hazards, such as water and confined space rescue, you may want to begin by forming a team in only one of these disciplines, become proficient in it, and then expand to a second discipline. Be careful not to start off trying to do everything at once; it is better to try to establish your proficiency in the most important areas first and expand later as you build on the initial capability and after you have developed your skills in this area.

The specialties you will cover with your team and the needs of your jurisdiction will help you to formulate a mission statement for the team. The mission statement is important because it will give direction and focus to a new team.

1. Establish a planning committee to develop a plan
2. Determine current capabilities
3. Prepare a concept of operations
4. Determine program management structure
5. Develop a staffing plan
6. Identify initial equipment and vehicle require-
7. Identify training requirements
8. Consider a plan for delivering recurring training
9. Estimate cost of team and develop a budget
10. Obtain management support
11. Obtain political support
12. Look for partnerships

### PHASE II: PLANNING

Once you have determined the type of team you need, you should develop a specific plan of action for creating the team. This plan should cover all aspects of team development including personnel, equipment, training. The most important planning steps are outlined in this section.

### 1. Establish a Planning Committee to Develop a Plan

Select a committee or team to develop your plan and appoint a chairman. The development committee should contain competent planners as well as individuals who might become the team leaders of the technical rescue team during its development and operation phases. Three to six people should be sufficient. You may want to place certain individuals that already have rescue experience or other related experience on the planning team. Define the goals for a technical rescue team development committee. What is the committee's charter? What are the objectives and parameters? When do they need to complete their planning? Make sure the committee understands the goals and that the goals are focused. A time frame should be given for the team to complete a plan. At least one chief or member of the department's top management team should be a part of the committee to help give it direction and to verify that it stays on course. The plan should address resources and operations for the following areas:

- **Personnel and Staffing.** Who will be the team leader(s)? What types of skills will be necessary to join the team? What will be the size of the team? (Refer to Chapter 5).
- **Equipment.** What equipment will be needed? What equipment do the individuals provide, what does the team provide?
- **Vehicles.** What type(s) of vehicle(s) will best serve your area and rescue mission?
- **Training.** What initial training and recurring proficiency training will be needed? (Refer to Chapter 7).
- **Administrative support.** Who will maintain records, equipment inventories, and provide program oversight?
- **Political support.** Will you need to obtain this or do you already have support from local leaders?

### 2. Determine Current Capabilities

Identify what equipment and training your department already possesses. Some of the equipment you will need is probably already carried on an engine, ladder truck, or rescue squad. Additionally, some of your firefighters may have already taken rescue classes on their own.

The more capabilities you can identify that you already have, the faster and cheaper it will be to start a team.

## 3. Prepare a Concept of Operations

Develop a basic concept of operations and a set of operational procedures.

The concept of operations will assist you in thinking through how you intend to operate and what resources you will need. It will also help you sell the program to management and the public by forcing you to think through how the team will be used.

An outline of the operational procedures are needed early in the process to demonstrate to management that you have thought through the program and have not left anything out. You can fill in the detail procedures when you get closer to putting the team into service. Development of standard operating procedures (SOPs) is discussed later in this section under *Phase IV*. Sample SOPs are contained in Appendix B.

## 4. Determine Program Management Structure

An organization considering the formation of a technical rescue team should identify and task personnel to address the fundamental requirements of the program. These personnel would comprise the program management team. A senior person should be identified as the senior program officer. This individual is the central administrator who coordinates all ongoing program responsibilities (i.e., scheduling meetings, developing proposals and correspondence, assigning tasks, tracking accomplishments, etc.).

Most departments have found it necessary to assign at least one rescue training officer to each shift. This position is responsible for the myriad issues involved in developing, conducting, and tracking training certification.

Likewise, the assignment of an equipment officer (or technician) is extremely important. It may be necessary to have one assigned on each shift or at each tech rescue station. These positions address issues related to equipment research and procurement, reception of new equipment, organization of the equipment cache, and ensuring that a maintenance and exercise program is addressed for all tools, supplies, and equipment on a recurring basis (weekly, monthly quarterly, etc.).

Due to the significant amount of development and staff work required when initiating a new program, the assignment of a staff/scribe position is quite beneficial. This person should be competent in computer software applications such as word processing, database and spreadsheet programs. Tracking information related to equipment and personnel details is made more manageable with the assistance of a computer.

## 5. Develop a Staffing Plan

One of the most critical development steps to accomplish in the formation of a new technical rescue team is to determine how many people are needed for your team. In general, staffing requirements must address filling all identified command/management staff as well as addressing the minimum number of personnel to effectively and safely conduct tactical operations. Staffing size will depend on the type of rescue team; a trench rescue team will need more personnel than a dive rescue team.

In general, all major technical rescue disciplines are staffing intensive, at least during the initial start-up phase of operation. Trench rescue and structural collapse operations may be the most intensive, easily requiring at least four or five specialists, overseen by command positions and assisted by non-certified personnel. Advanced rope operations may require a sizeable cadre of personnel for raising operations. The majority of personnel operating raising or belay lines need not be certified personnel (but must be under direct control of certified personnel).

The staffing plan should also address the number of personnel required per rescue unit (vehicle). Many departments staff heavy rescue squads or other specialized units to address specific tactical requirements. Other departments may not be able to afford this luxury due to size limitations or other restrictions. Many departments already have minimum staffing requirements of three, four, or more personnel required on each in-service unit. In career departments, local charters or union regulations may impact these decisions. Additionally, if rescue personnel must be drawn from multiple stations, the plan should discuss how the station positions will be filled while rescue team members are on a call. The staffing plan should also address whether the team will be comprised of personnel from career ranks only, volunteer ranks only, or from both. Additional discussion of team staffing is provided in Chapter 5.

## 6. Identify Initial Equipment and Vehicle Requirements

An analysis of the equipment needs should be conducted separately for each discipline. Then the separate lists can be combined into a single equipment procurement list. Most fire departments or rescue agencies may already possess much of the identified equipment. In this case, it may only be necessary to gather the equipment in a central location or develop a resource list denoting each item's location and a mechanism to gather it for response use. This process may dramatically reduce the funds needed to procure all necessary equipment for the team's operations, however it will require time in an emergency to gather the equipment if it is not kept at a central location.

Some departments have sent one member to training classes to learn what rescue tools are necessary for a new team. This is an excellent way to establish basic knowledge of equipment capabilities, which is important for identifying what is needed.

In most cases, if funds are limited, the purchase of equipment could be prioritized based on the greatest need for one or more of the identified team disciplines. Purchases that increase personnel safety should receive higher priority, while purchases that expand capabilities should be a secondary priority. In any case, safety and the need for a certain amount of redundancy in equipment must be stressed. Obviously, if a key tool or piece of equipment malfunctions, or is unavailable due to maintenance, the ability of the team may be critically impaired.

It may be easiest to request copies of equipment lists from technical rescue teams that are in place and use one or more of these as a starting point for the equipment cache development. The U.S. Fire Administration's *Technical Rescue Technology Assessment* report contains information about equipment needed for technical rescue, and is available free (refer to Appendix C for information

on how to obtain this document). More information on rescue equipment is contained in Chapter 8.

Once you have determined what equipment is necessary for the team, you can consider what vehicles are capable of carrying the equipment and team members. You may be able to fit the equipment on an existing unit, or you may need to purchase a new rig. Some teams use a cargo trailer, convert an old unit, or request a vehicle be donated by a local business.

## 7. Identify Training Requirements

The training to competently and safely address each individual capability is intensive. The greater the number of specialties a technical rescue team assumes responsibility for, the more difficult is the task of bringing personnel up to the necessary training and skill levels.

In the planning stage, you must identify what training you will need and what training is available. Training needs will be determined by the team's focus. They will also be determined by any local or state training requirements (this is particularly important in states regulated by their department of occupational safety and health). When will the training be delivered? Who will deliver the training?

Additional information about technical rescue training can be found in Chapter 7.

## 8. Consider a Plan for Delivering Recurring Training

Maintenance of skills is critical to the competency of rescue team members. Develop a plan which establishes minimum continuing education standards for members. Some of the recurring training can be done on an individual basis, but the entire team should convene for a team training session several times a year. Check with your state training agency to see if it has already established continuing education requirements for rescue team members. The cost of recurring training must also be considered.

## 9. Estimate Cost of Team and Develop a Budget

Preparing a cost estimation for the team is time consuming and requires research, but it is a very important step in the development of a team.

The city manager or board of supervisors will want to see a detailed budget plan before approving a team.

The first step in planning a budget is to list separately the major types of rescue you plan to undertake (i.e. water rescue, confined space rescue, trench rescue, etc,). Consider each of these as an individual heading. Under each area, list the training, equipment, and apparatus you anticipate needing to start the team. List all the equipment and training you would like to have - do not leave anything out. Costs for each of the following areas must be considered:

- *Personnel hours*
- *Training and continuing education*
- *Texts and materials*
- *Travel expenses*
- *Equipment*
- *Vehicles*
- *Protective Clothing*

Next, obtain at least two estimates of the cost for each item on each list (estimated costs for some equipment is listed in Appendix D). This step requires heavy research. Do not just rely on costs in a catalog. Thorough research on pricing involves talking with manufacturers or distributors to find out product capabilities and limitations so that you can compare different products. You also may be able to discuss special pricing. Round prices up so that you do not end up under-budgeting.

Once you have completed pricing and product research, compare the different products and prices to determine what is best for your needs. Total the cost of each training, equipment, and apparatus item to determine the maximum start up cost. Items that are not immediately essential to initiating a team may be eliminated and budgeted in the future. This will help lower your start up costs. You must determine, however, what items are absolutely essential to begin a team. The total of the cost of the essential items is the minimum start up cost. (Refer to Chapter 4 for more information on obtaining team funding.)

## 10. Obtain Management Support

This is probably the most important step in developing a technical rescue team! You must sell the program to your fire department management. Battalion Chiefs, Assistant Chiefs, Deputy Chiefs, the Fire Chief, the City Manager, and in the case of volunteer departments the Board of Directors, all have to recognize the benefits of this kind of program and support it if you are going to succeed. Is the program technically feasible? Get all of your supporting materials ready and rehearse them before going public with the plan. You may only get one chance to sell your program and you can assume that some of the audience will not be favorable or supportive. Be ready for them. Cite other departments in your region or state that have teams and summarize how their teams are working for them.

Your objective in this step is to get permission to develop the Technical Rescue Team. Obtain support of the fire chief first, and then present the team concept to the City Manager. Usually management will want time to think over the idea. Try to set a specific date for a decision or another meeting where a decision will be made.

If your fire department operates independently of any outside jurisdictional oversight, you can minimize this step. However, if you don't know how your Chief feels about a rescue team, don't assume he will buy into it without significant convincing.

## 11. Obtain Political Support

Develop a plan to obtain political support. This is necessary to secure funding for this program. You will need political support to get funding if your department does not have an independent funding source. Remember that eventually you or someone else will have to go to the department board of directors, the town council, or the county board of supervisors to procure funds for the project.

Be prepared to answer questions about the team. Common questions asked by management and elected officials include:

"Why do we need a technical rescue team?"

"Don't we already have those capabilities?"

"How much will this endeavor cost?"

"Can't we get rescue services from other jurisdictions?"

"Can we share the costs of a team with another jurisdiction?"

| Worksheet to Estimate Initial Budget for Technical Rescue Team | | | | | |
|---|---|---|---|---|---|
| Expense Area | Estimate No. 1 | Estimate No. 2 | Estimate No. 3 | Check If Essential for Team Startup | Best Estimate |
| **PERSONNEL** | | | | | |
| Research, Planning and Development Time | | | | | |
| Training Wages for Rescuers | | | | | |
| Replacement Staff Wages (during emergencies or during training | | | | | |
| Retraining or Continuing Education Wages | | | | | |
| Ongoing Management of Team | | | | | |
| **TRAINING CLASSES** | | | | | |
| Classes | | | | | |
| Travel Expenses | | | | | |
| Texts/Materials | | | | | |
| **EQUIPMENT** | | | | | |
| Tools and Appliances | | | | | |
| Equipment Storage Facilities | | | | | |
| Maintenance | | | | | |
| Equipment Replacement | | | | | |
| Protective Clothing and Team Apparel | | | | | |
| **VEHICLES** | | | | | |
| Purchase or Reconditioning | | | | | |
| Insurance & Maintenance | | | | | |
| **TOTAL ESTIMATE** | | | | | |

**"How often will this team be used?"**

**"Do we really need a team for rescues that happen so infrequently?"**

If you have gone through each of the previous steps and done your homework, you will be prepared to answer questions like these. Be ready to make specific, concise points to justify your request for approval of a new team. Below is a list of tips that may help you win political support.

• Discuss your concept of a team with individual elected officials before presenting it to the entire body of officials. Be sure to have support from the county manager or mayor before going to the elected officials.

• Prepare a list of hazards in your response area and note the dangers and risks associated with each. Give this to the elected officials. Be sure to note the risks presented by each to both citizens and rescuers. Discuss what will be the acceptable risk thresholds.

• Create a video or slide presentation that will demonstrate the hazards that exist in your area. You also may gather action pictures of rescue teams already formed to demonstrate team capabilities.

• Have charts prepared that demonstrate the need for a team and show the number of rescue incidents you have run in the past and you expect to run in the future.

• Have charts prepared that outline a plan for developing the team.

• Be prepared to discuss regulations, such as those for confined spaces, which may require you to train your personnel to a certain rescue level in order to make certain rescues. This alone may justify the team. (Refer to Chapter 6 for information of laws and standards affecting rescuers.)

• Become familiar with other rescue programs around your region or state that will serve as examples.

## 12. Look for Partnerships

Partnerships are especially helpful to new teams because they can assist with funding initial costs. Local industry may have confined spaces and, under OSHA regulations, may be required to have a confined space rescue team. The local industry, however, may not have the personnel necessary to have a team, and may request assistance from the fire department to serve as their confined space rescue team. In exchange, the fire department receives funding from the industry to pay for training and equipment. The community also benefits from the availability of a confined space rescue team within the fire department.

This step in the development process could occur before obtaining management approval, but often the management wants to know about plans for a technical rescue team before the fire department goes out into the community to build partnerships. Therefore, this step may need to fall after management approval is obtained.

## PHASE III: DEVELOPMENT OF TEAM

1. Select the team members
2. Train the team
3. Purchase equipment and uniforms
4. Purchase vehicles
5. Provide administrative support

### 1. Select the Team Members

One of the best methods for selecting team members is to conduct interviews. Start by soliciting personnel who are interested in joining the team. Have them complete a short questionnaire about why they want to join the team and what skills they could bring to the team. Any person who has outside skills in areas such as construction, rappelling, diving, etc. will bring added skills at no extra cost to the department.

It is imperative that you clearly delineate what additional demands and responsibilities will be expected of those joining the team before they officially join. For instance, they may be expected to participate in continuing rescue training in addition to continuing firefighter training. In a volunteer organization, it is especially important to delineate expectations in advance because technical rescue team demands will probably take more time than just firefighter demands.

Another consideration when selecting a team is to recruit members who have emergency medical training. Many rescues will require personnel to perform technical rescue team and emergency medical skills. Chapter 5 provides additional discussion about selection of team personnel.

### 2. Train the Team

The team will need a thorough initial training program on all the equipment and the rescue techniques. Train your people to handle the specific target hazards in your response area. Ensure that the training program includes a mix of hands-on and technical classroom topics. Realistic training scenarios will require working with area contractors or other organizations to donate trenches, buildings, or other facilities. Refer to Chapter 7 for additional information on rescue training.

### 3. Purchase Equipment and Uniforms

Purchase the equipment your team will need based on its mission objectives and based on equipment needs you have previously defined. Start with the basic equipment and add the more complex technical rescue equipment as you proceed. Leave the "nice-to-have" equipment for purchase after your initial training is complete and after you have gained the experience to evaluate what tools you really need to add. (Note that you may have to purchase some equipment before team members begin training because they may need the equipment during class.) Refer to Chapter 8 for more information on equipment and uniforms.

### 4. Purchase Vehicles

In the planning phase you specified the general type of vehicle you would need (trailer, four wheel drive, etc.). In this step, detail vehicle plans are necessary including equipment storage layout to make ensure that equipment will fit in the vehicle. You should allow about a one-third growth factor for future equipment additions. Make sure you have a secure storage area for everything to avoid damage or injury. If a trailer vehicle is planned, verify that the trailer hitch is sufficient to handle the weight of the trailer and equipment.

### 5. Provide Administrative Support

One part of the planning process of technical rescue team development which is usually forgotten is the administrative effort necessary to get the team started. Members of the team or support staff should be recruited to maintain the records for the team. Example record keeping tasks include:

- *Team Roster*
- *Immunization and Health Records*
- *Call Back Lists*
- *Equipment Inventories*
- *Equipment Repair/Maintenance Records*
- *Team Activation Checklists*
- *Training Records*
- *Training Schedules*
- *Expenses*

The counterpart of identifying and developing a recurrent training program is addressing the tracking equipment and accounting of team member attendance at training. This is an important administrative step. Additionally, you must track all expenses related to training and equipment. This information will help you conform to your budget and will be necessary for reporting to city administrators and elected officials.

## PHASE lv: DEVELOPMENT OF STANDARD OPERATING PROCEDURES

| 1. Obtain or write adminis trative and operational SOPs for the team |
| 2. Review and revise SOPs regularly |

Standard operating procedures (SOPs) are an integral part of a technical rescue team. Some fire departments and rescue teams choose to function without SOPs, but SOPs are vital to have a safe and organized rescue operation. SOPs establish technical rescue team organization, processes, and techniques before an emergency incident occurs. SOPs should answer questions such as who is in charge, what equipment will be used, what techniques will be used, who is qualified to perform a technique, what is expected of each responding unit, and what staffing is required at a rescue incident. Most importantly, they provide a structure by which a technical rescue team can respond in an organized fashion to the chaos and uncertainty presented at almost any emergency incident.

Many technical rescue teams have already developed SOPs which may serve as a guide for the development of ones for your department. Sample SOPs are included in Appendix B. Obtain copies of several SOPs for the types of teams you plan to develop and review them in detail. Use those features that you feel best apply to your goals and situation. The regulations and standards discussed in Chapter 6 may provide assistance in developing SOPs.

Technical rescue teams should consider forming two types of SOPs: administrative and operational. The procedures should be consolidated into one manual, and they should be fully integrated with the fire department's existing SOP system. Administrative SOPs provide the framework for the personnel structure of the team. Operational SOPs describe things such as techniques and unit responsibilities that are used at an emergency incident. Each of these is discussed further below.

### 1. Obtain or write administrative and operational SOPs for the team.

ADMINISTRATIVE SOPs

The administrative section should address:

Chain of command. The administrative and operational sides of the chain of command for the technical rescue team should be clearly defied.

**Specialty certification requirements.** The tactical capabilities that the team is responsible for must be clearly identified. The training requirements related to each discipline must be fully defined. This should include the initial training required for certification in each discipline, as well as continuing education requirements.

**Unit/equipment requirements.** This section would define the types of vehicles and equipment for the technical rescue team. Any requirements related to the management, organization, and maintenance of the team equipment cache must be addressed. This should include the development of a routine cache maintenance/exercise schedule to ensure the operational readiness of all tools, equipment, and supplies.

**Unit staffing.** The staffing of specialty vehicles, if dedicated, should be identified. This would include any minimum staffing requirements, if mandated. Or, it may only be necessary to mandate the number of specialty personnel required to effectively handle technical rescue operations (the number may vary by incident type). In any case, the number of certified personnel and/or minimum staffing requirements should be clearly understood by all.

OPERATIONAL SOPs

The operations section should address:

**General operating procedures.** This would cover the types of incidents the team is responsible for, the dispatch of standard/specialty units for any type incident, and general or first responder actions (i.e., standards for non-specialty personnel) to be taken upon arrival.

**Incident-specific operating procedures.** A general overview of the tactical operating procedures should be defined. These may be separated by event type (i.e., trench, structural collapse, rope, etc.), if necessary Unique requirements or considerations for each discipline should be addressed.

**Regulations/requirements.** Certain technical rescue operations are impacted by local, state, or Federal regulations. These regulations should be included in your procedures.

**Scene management procedures.** Most departments already have an incident command system already in place. The basic command structure can be used at any incident, including a technical rescue incident, but additional technical rescue command positions should be added to it. This section of the SOPs must detail how technical rescue incidents will be commanded. A command organizational structure designed for technical incidents should be prepared. Refer to Chapter 9 for more information on technical rescue incident command.

**Tactical command worksheets.** Most departments with technical rescue teams have developed some type of tactical checklist or command worksheets to assist technical rescue command personnel in the management of an incident (Figure 3-2). These may be developed for each discipline, if necessary.

**Team Activation.** Activation procedures must be developed and exercised by the full team to ensure their completeness and adequacy. These procedures should cover: team callout, staging areas, equipment movement to the staging area, food procurement if required, list of personnel actually deployed and family contacts, daily status reporting to the department if the team is deployed away from home, and other related lists.

## 2. Review and Revise SOPs Regularly

SOPs should be reviewed by a group of team members on a regular basis (at least annually) to ensure that the procedures are up-to-date and meet the needs of the team. In addition, after a major rescue incident, the procedures should be reviewed and revised if they proved to be faulty or inadequate.

*Figure 3-2. Fairfax County (VA) Five and Rescue Department Technical Rescue Tactical Worksheet*

## FAIRFAX COUNTY FIRE AND RESCUE DEPARTMENT

### Technical Rescue Incident-Tactical Worksheet

☐ Technical Rescue officers begin size up of situation

☐ Other personnel ensure firm scene control is established
- ☐ Hazard/working area cordons set up
- ☐ Bystanders/coworkers/nonessential personnel removed from cordoned areas
- ☐ Utility/personnel hazards Identified/mitigated

☐ Have Technical Rescue officers Identify plan of action

☐ Conduct short team briefing
- ☐ Sketch of operation (use erasable marker board on trailer/legal pad)
- ☐ Identify procedures to be used
- ☐ Identify equipment that will be required for operation

☐ Appoint Technical Rescue command officers (and Issue vests)
- ☐ **Rescue Leader**   ☐ **Rescue Safety**
- ☐ **Rescue Equipment**   ☐ **Rescue Personnel**

☐ Consider patient removal operation early
- ☐ Assign victim removal officer to Identify/coordinate operation
- ☐ Removal tactics/personnel should be preplanned
- ☐ Necessary equipment should be Identified/assembled

☐ Ensure that equipment sector Is established adjacent to working area
- ☐ Anticipated equipment should be assembled
- ☐ Equipment should be checked prior to use
- ☐ Ensure equipment Is inventoried/tracked throughout Incident

☐ Ensure personnel sector Is established
- ☐ Non-Tech Rescue personnel should be staged at this area for assignment
- ☐ Unassigned Tech Rescue personnel staged at this area

☐ Number of personnel In hazard and working areas should be limited
- ☐ Personnel officer should ensure unnecessary personnel report back to Personnel sector

☐ **Trench Incidents**
- ☐ Ensure ground pads placed at trench edges
- ☐ No one enters unshored trenches deeper than 5 feet

☐ **Structural Collapse Incidents**
- ☐ Ensure full structure is reconned for victims
- ☐ Request Building inspector?

☐ **Confined Space Incidents**
- ☐ Ensure that Confined Space Entry Checklist Is completed prior to operations
- ☐ Ensure that back-up rescue team Is ready prior to committing entry team
- ☐ Request Light/Air Unit? (supplied air system can be run off Light/Air Unit)
- ☐ Ensure all 1200 feet of supply air hose Is laid out prior to entry (300 feet/person)

☐ **Advanced Rope Operations**
- ☐ Ensure rope systems are double checked prior to use (Rescue Leader/Rescue Safety)
- ☐ Ensure helmets are worn by personnel over edge/down below
- ☐ Rope system movement commands to be called out by Rescue Leader
- ☐ Rescue frame needed?
- ☐ Rope system hand winch needed?

*Courtesy of Fairfax County Fire and Rescue Department*

# CHAPTER 4: FUNDING REQUIREMENTS AND POTENTIAL SOURCES

**T**echnical rescue operations can be an expensive undertaking for many jurisdictions. Given financial constraints, locating funding sources can be one of the most difficult hurdles to overcome for new rescue teams. Existing teams often fight for their budgets each fiscal year and are always looking for new and creative ways to finance their operations.

This chapter discusses where the money goes when forming a team, sources of funding, and ideas for justifying a team's expenses.

## THE FINANCIAL COSTS: WHERE THE MONEY GOES

To help establish the type of rescue service needed in the community and the financial support the community is willing to give, it will be important to understand where the money will be spent and how much money will be needed. Appendix D lists estimated costs for equipping a team in various rescue disciplines. These are rough estimates to give program managers, team leaders, and administrators an idea of the financial commitment that may be necessary. Chapter 3 contains a worksheet to help you estimate a team's expenses.

**Initial Training.** Training costs can range from several hundred to several thousand dollars per student per course. Shortcuts should not be taken with training funds. Thorough training is necessary to have a safe and effective rescue capability.

You may consider training team members over a two or three year period to spread out the costs. Budget for personnel to receive basic awareness level training the first year and operations training the second year. A few select members could later be trained to the technician level or higher. Efforts should be made to have incident commanders participate in training, so they have an understanding of the rescue operations and equipment. This will also help when commanders develop standard operating procedures for their rescue teams.

**Continuing Education.** Funding for technical rescue teams must take into account a commitment to continually train and retrain personnel. It is not enough to initially train and equip a team; to be effective members must constantly practice their skills and learn new ones. For example, it has been estimated that proficiency at technical rope rescue skills is reduced within six months after completing a rope rescue course if training is not maintained. Continuing education for technical rescue may be even more important because rescue incidents are usually rare, unlike fire or emergency medical incidents.

Continuing education expenses are incurred from sending personnel to refresher courses or advanced courses which count toward recertification, or from holding a special continuing education drill. Holding a drill is generally the cheapest alternative but in most cases it will not provide certification for attendees. Legal mandates may require regular recertification training, which can be a more expensive proposition requiring you to hire an instructor that can recertify personnel.

Career departments may need to allocate payroll funds for training that cannot be done during on-duty schedules. Forty-eight hours of continuing education per year is not an unreasonable requirement. Can your department afford to conduct this amount of training during normal hours, or will overtime be involved?

**Equipment.** Equipment costs will depend on the type of rescue capability the community desires. Basic equipment to perform many rescues such as rope, ladders, and breathing apparatus may already be available on existing emergency apparatus. In many cases, supplemental equipment to augment your rescue capabilities could be purchased for several thousand dollars. Advanced capabilities, however, generally require expensive specialized equipment.

Costs for equipment storage and maintenance must be considered also. Large caches of equipment must be kept secured but accessible in an emergency.

**Transport Vehicles.** The major vehicle expenses a rescue team will encounter are for purchasing or retrofitting, maintenance, and fuel. The amount of money spent on vehicles to transport a team and its equipment will also vary widely, from no cost to add additional equipment on existing vehicles, to $300,000 or more for a dedicated heavy rescue apparatus. Vehicles range in type from pickup trucks and sport utilities to box vans and heavy squads. Many teams pull gear in trailers.

Opportunity exists for having vehicles donated. Many utility companies donate vans or trucks to non-profit agencies; your department may be eligible. Private companies have donated beverage trucks or tractor trailer boxes to teams. Using these local resources can reduce your costs. Annual maintenance costs must also be accounted for, especially if an additional unit is added to a fleet of apparatus.

**Insurance.** The cost of insurance is often overlooked. You may need to purchase insurance for equipment, vehicles, personnel, or malpractice. A team formed by a fire department or group of departments may be able to absorb insurance costs into the department's existing policy. In this case, you must verify that existing policies will extend coverage for these new operations. A fire department may need to add or make changes to its insurance policy to make sure its members are covered for confined space rescue or water rescues - duties which may not be listed in the department's charter, by-laws, mission statement, or articles of incorporation. Local officials and attorneys should be involved in this process. Insurance issues for consolidated teams, mutual aid coverage, and out of jurisdiction training also must be addressed (refer to Chapter 10).

## JUSTIFYING EXPENSES

City or county administrators will want you to justify the expenses necessary to start and fund a rescue team. A team may be easier to justify in a community with a large risk potential; smaller or less frequent risks make justifying funding more difficult. The expenses must be justified to the many individuals who control the financing; attempts should be made to involve all of them in the program development for the team. A team leader must justify the funding to a fire chief, who must justify it to the city manager, who may have to justify it to the city council. Today, public budgets are placed under a microscope - a clearly defined mission for a team is as important to its financial success as to its operational success.

Linking funding requests to existing local needs - especially past incidents and safety concerns, as discussed in Chapter 3 - provides more legitimate justification of the funding requests. Local, state or Federal regulations can also be used to justify a

team's expenses. A review of state OSHA regulations and other rescue standards should be conducted. Team leaders can use mandates such as OSHA's Permit-Required Confined Space standard to justify team expenses. All of the individuals involved in the decision making process should understand that, unlike most fireground operations, the agencies involved in technical rescue may be subject to severe fines and sanctions if they fail to comply with established OSHA standards while performing their duties. Potential lawsuits brought by the public for failure to provide service may also be part of the overall equation with regard to financial decisions.

The department and administrators must carefully evaluate what the public expects of the fire department. Administrators may feel the risk potential is low and may not want to spend any money on a team. It is important to let them know what types of rescues you are capable and not capable of safely performing. The managers may not realize that special rescue capabilities are not available as part of your normal service. Too many would-be rescuers have died attempting to perform rescues they were not trained or equipped to handle. Making the public and government administrators aware of these issues may help justify the team. The administrators in control of the money must be made aware that their decision may determine whether the department will or will not be able to make certain rescues.

The key is to be prepared to justify the need and costs for a team. Refer to the section on obtaining political support in Chapter 3 for additional ideas about how to present your rescue team plan to managers and elected officials.

## FUNDING SOURCES

Finances for a technical rescue program may come from many different sources. Often, municipal tax funds are allocated to add technical rescue services to existing emergency service providers. Donated money and equipment can also be used. State grants may be difficult to secure but may provide the necessary seed money to get a program established.

Funding methods are limited by the effort, imagination, and regulations of the team and its parent agency. A good source of information about various methods of funding is the United States Fire Administration's *A Guide to Funding Alternatives for Fire and Emergency Medical Service Departments,* which is available for free from the USFA (see Appendix C).

**Direct Funding from Local and State Government.** Direct funding is where a governmental organization budgets a designated amount of money toward the technical rescue team's projected operational costs. These funds are usually tax revenues from the government's general fund or from the department's annual budget. While technical rescue teams may expect this direct support, securing the initial funding is the exception, rather than the rule.

**Cost Sharing:** Multiple Department Funding. Many communities with limited financial resources may be able to establish technical rescue capabilities by entering into agreements with surrounding jurisdictions. This helps to promote economies of scale as team leaders attempt to justify their existence despite a limited number of responses and high costs. In this way the financial burden for technical rescue is divided among the different communities. Areas with a need for different types of rescue may choose to divide up the responsibility, with one department developing a trench or collapse rescue ability, another department developing a water rescue or dive rescue ability, and one department establishing rope rescue ability. Each jurisdiction agrees to provide its specialized services to the other ones through automatic mutual aid responses.

Another method to reduce costs through consolidation of resources may be to form an interjurisdictional team with either a single or multi-rescue capability. In this arrangement, interested members of each department could join the team, and each department could share the financial costs for training and equipment.

The funding mechanisms for existing regional hazardous materials teams around the country may serve as examples of how jurisdictions can share costs for technical rescue. For example, the Southegan Mutual Aid Response Team (SMART), a hazardous materials team is an example. This region of New Hampshire conducted a hazard, resource, and risk assessment for its participating communities, and distributed the costs of the team to the various communities based on the risks and resources in each area.

**Public-Private Partnerships.** Funding directly from the local department or jurisdiction is not the only source of funding that should be explored. Funding assistance can also be provided by local industries by forming a partnership. Public-private partnerships are a popular, economical, and efficient way of providing rescue service. This funding concept is discussed further in Chapter 3.

The Tidewater Virginia Regional Heavy and Tactical Rescue Team is an example of a public-private partnership. The team receives operational funding from regional utilities, building contractors, professional and business organizations as well as municipal budget provisions. Montgomery County, Maryland has received donations from the Associated Builders & Contractors (ABC) and the Washington Gas utility company in the form of equipment and dollars for some of its specialty teams.

**Local Clubs and Community Organizations Funding.** Funding can be acquired from fraternal and community oriented organizations. The United Way and local civic groups often support special projects with funding. It is as important to explain the team's mission to these organizations as it is to explain it to the politicians. Often, your customers are members of these organizations. Recognition by a community organization merely amplifies and extends the recognition of the technical rescue team's mission, and helps illustrate financial needs. It is also important to provide recognition back to the community groups that provide direct or indirect support to the organization.

Some organizations have painted contributors names on their equipment or vehicles. A rescue boat at one fire department is named for a family that donated funds to facilitate its purchase. The Philadelphia Fire Department advises community groups, business groups, or corporate sponsors when their donated equipment was used for a high profile incident that saved a life. Technical rescue teams should share the glory of a mission with their supporters.

**Seized or Confiscated Property.** Another avenue that can be explored are local and state law enforcement confiscation programs. Many jurisdictions confiscate vehicles and trucks, as well as equipment, for various reasons such as drug interdiction. A number of departments across the United States have been able to take advantage of this resource.

**Excess and Surplus Government Property.**
The Federal government has both surplus and excess property programs which may provide another avenue to find tools and equipment. These can usually be accessed through the respective state emergency management agencies or directly from the Federal government. The Federal government routinely releases equipment and supplies it no longer needs to other governmental entities for use at no cost, other than shipping. This is termed excess property. There are certain requirements with regard to obtaining excess property. The recipient must maintain an accountability system and report damage or destruction of nonexpendable property to the General Services Administration on an annual basis.

Surplus property indicates that equipment and supplies have gone through the Excess Property System and that no Federal agency has acquired these items. Usually, this equipment is assigned a condition class and must be closely screened for practicality and suitability by the requester. Local departments can access Federal surplus property through their own state surplus donations program. There is usually an administrative charge by the state for acquiring such items.

Two publications may assist in finding excess surplus property. First is the *Personal Property Utilization and Disposal Guide* published by the General Services Administration (see Appendix C). This 40-page pamphlet contains information about excess property, terms and definitions, processes for acquisition, and addresses and telephone numbers of the Federal Supply Service Bureau regional offices. The second publication is *How to Buy Surplus Personal Property (from the United States Department of Defense)*. This pamphlet contains similar information regarding the acquisition of DoD surplus property. It can be obtained from the Defense Reutilization and Marketing Service (see Appendix C).

**Grants.** Your state may offer grants to assist with the purchase of fire department equipment and vehicles. Some grants have restrictions as to what type of agency may apply. Grant funds are generally available for very specific reasons. Private grants from local industry or foundations may be available for rescue needs in a community, such as the purchase of a specific rescue tool, equipment, or a vehicle. The process of applying for and receiving grant funds can be complicated, and thorough research into the process should be conducted by team members to avoid wasting effort.

Some states or local jurisdictions have dedicated grant funding mechanisms for fire and EMS that may be applicable to technical rescue. The State of Maryland, for example, has a grant program that has funded rescue apparatus and equipment for local fire departments. In past years, the Federal government has provided grants for rescue team development. Thorough investigation of state and Federal grants can pay off, but remember that the grant proposals and requisition process can be a very time consuming process.

**User Fees and Cost Recovery.** User fees are essentially charges for services rendered. Many departments have user fees for services such as plans review, inspections, EMS transport, and even fireground operations. Hazardous materials teams have been charging user fees for service for many years, aided by Federal legislation which allows them to recover expenses incurred during hazardous materials operations. Rescue teams may be able to charge similar fees for rescue operations. Several mountain rescue agencies charge these fees for rescuing stranded climbers. The San Antonio, Texas Fire Department charges $400 per water rescue. The fee was instituted both to raise funds for special equipment and training, and to deter citizens from taking needless risks. Most private medevac helicopter services and many public helicopter services charge fees for performing rescues or for transporting patients to hospitals. The United States Coast Guard charges fees for certain rescues, as does the US. Park Police. Rescue teams may be able to retrieve funds expended during rescues by billing a third party such as an insurance company or construction contractor. A trench incident may result in EMS fees billed to the victim, and trench rescue fees (for equipment use, personnel costs, etc.) billed to the contractor.

**Permit Fees.** Some funding may be raised through charging permit fees to construction contractors or businesses performing work that could require technical rescue. Contractors generally must to apply for local building or zoning permits in the locality were they are conducting work. A fee or additional permit could be added if the work involves confined spaces, trenches, or other hazardous operations. The contractor could be required to pay a small fee for the permit or could be required to take out some kind of insurance which would ensure the rescue agency will be reimbursed should a rescue be necessary. Local and state ordinances may be required to institute such fees. A permit program offers the additional benefit of allowing the rescue team the ability to conduct a pre-incident plan for a known hazard.

Training Fee. Opportunities exist for raising funds through rescue training. Training courses could be provided for rescuers from other agencies who want to achieve competency in certain rescue disciplines. Many teams currently conduct this kind of program. Other training fees may come from private companies and building contractors who are required to have personnel trained for operating in confined spaces, in trenches, or on rope. Safety regulations often require these companies to have personnel trained in rescue or first aid on site. Fees can be charged to train these personnel. This will raise money and also help to reduce accidents by increasing the competency of workers in your area. It will be important to address liability issues when exploring the idea of establishing a training program for groups outside of your department.

# CHAPTER 5: PERSONNEL AND STAFFING

**T**he backbone of a good technical rescue team is well trained, experienced personnel. The personnel can be either career or volunteer firefighters, or come from other backgrounds. The success of a team will be influenced in part by the personnel selected for the team and their ability to function together as a team. This chapter discusses many of the personnel and staffing considerations necessary when forming a rescue team.

## TYPE OF PERSONNEL NECESSARY FOR A TECHNICAL RESCUE TEAM

In most fire and rescue organizations certain personnel naturally gravitate towards technical rescue programs. The capabilities required for personnel on a technical rescue team often involve a high degree of mechanical aptitude and physical strength. Individuals who are skilled working with their hands and who exhibit ingenuity, resourcefulness, and inventiveness are valuable. Trade skills (i.e., carpentry, plumbing, electrical, metal work, electronics, heavy equipment operators, etc.) can be extremely useful and pertinent. Individuals with special skills or training can bring their talents to a team at no additional cost to the department. Carpenters may have the knowledge about how to build shoring. Construction workers may be familiar with heavy equipment operations. Civil engineers may have knowledge about structural integrity during collapse operations. Recreational rappellers or kayakers may have skills for rope or water rescue. These qualifications should be assessed during the recruitment process.

Rescue team personnel must also be willing to meet the minimum standards required to achieve and maintain special training certifications. The standards may require that each member attend a certain number of training sessions on a yearly basis. Certain sessions may be legally mandated requiring attendance by all personnel.

## PERSONNEL PHYSICAL REQUIREMENTS

Due to the demanding physical aspects of technical rescue operations, it is apparent that the personnel comprising the team must be in good physical condition. Technical rescue personnel must be capable of performing functions such as handling, transporting, and setting up heavy equipment. In addition, good physical conditioning should reduce injuries.

Many departments have established medical requirements and physical fitness programs. The specific needs of specialty teams should be evaluated. NFPA *1582, Standard on Medical Requirements for Firefighters,* and NFPA 1500, *Standard on Fire Department Occupational Safety and Health Program,* may be used as references for further information (see Appendix C). If a department does not have a physical standards program in place, technical rescue program managers should seriously consider implementing one for their specialty team personnel.

## SELECTION OF PERSONNEL FOR TEAM

Personnel application and selection are an important component in the organization and development of a technical rescue team. The selection process should screen candidates for their commitment, consider previous rescue training and experience and skills learned outside of the department, as well as leadership, and physical capabilities.

Many teams begin the selection process by announcing the formation of the team and requesting letters of interest or resumes from interested individuals. Personnel comprising the team certainly need to be interested, motivated, and committed to the program. Departments may want to conduct written and/or oral interviews of candidate participants to ensure the candidates understand the commitment they are making and as a means to select the best qualified individuals. It also may require special physical agility testing, especially if this is not done when members join.

As part of the selection process, a department may require members to make a commitment to be a team member for a certain period of time. Some departments have required personnel to sign an agreement to remain on the team for a set period, such as five years. This can be justified in terms of the time, effort, and funding involved in training and maintaining the skills of the personnel on the team. This is a valuable commodity and investment. It is harder to require volunteer personnel to sign an agreement, although a volunteer department can create an agreement that requires a volunteer to repay the department for courses if the volunteer leaves within a certain period after completion of the courses.

Departments with a large pool of rescue team candidates can generally be more selective in terms of the physical condition of personnel they choose to be on a team. Smaller departments, however, may not be able to afford the loss of a firefighter's eligibility to be on a team due to physical standards. It is important to establish some type of physical requirements, but these departments may avoid very rigorous physical standards and instead ensure that team members do not have any risky physical condition such as a heart condition. All departments, regardless of size, should establish some sort of physical screening standard for technical rescue personnel.

Some departments may be forced to look outside of the agency for personnel to staff the team, especially if personnel with specific areas of expertise or skills are needed. Some agencies have hired individuals who have been active members of rescue teams in other departments to start up programs in their agencies.

## INCORPORATING FIREFIGHTERS, EMS PERSONNEL, AND NON-RESCUE PERSONNEL INTO RESCUE OPERATIONS

A dedicated technical rescue team must become an integral part of the overall department operations. Most teams are integrated with the suppression and emergency medical components of the fire department and are only activated when needed for special operations during emergency operations. Specially trained rescue personnel will direct operations, but generally they will need the assistance of non-specialty personnel, who can perform tasks that do not require special training.

This need implies that not only must the technical rescue team's operating procedures and team training address this aspect, but rescue training for all personnel departmentwide should be addressed. Some departments have developed a first-responder or awareness level of training for all department personnel that is based on a tiered response system similar to that discussed

in Chapter 1. This defines actions that should or should not be taken by non-specialty personnel initially arriving on the scene of a technical rescue incident. They usually arrive on the scene first, and they may be on the scene for a significant period of time prior to the arrival of the specialty team.

Effective scene management procedures should address this eventuality. All department personnel must be trained in scene safety, information collection, and hazard identification. All personnel should clearly understand technical rescue hazards and especially what not to do at the inception of an incident. Personnel must understand that they absolutely should not enter an unshored trench to begin rescue operations. Personnel must not enter a confined space without proper respiratory protection, ventilation, lighting, and back-up team support. Current OSHA regulations mandate other strategic and tactical requirements (i.e. atmospheric monitoring, back-up rescue team) prior to the entry of rescue personnel into a confined space. These are but several examples of incidents where first arriving non-specialty personnel must not commit to rescue operations without proper training and equipment.

The most effective way to address these requirements is through the development, training, and implementation of stringent scene management procedures. In general, these should address at least the following:

- *Actions to be taken or not to be taken by first arriving personnel*
- *Information collection/scene size-up*
- *Scene control (remove bystanders/erect cordons/etc.)*
- *Assessmen t/mitigation of hazards/utilities*
- *Command structure*

These actions set the stage for successful technical rescue operations (refer to the standard operating procedures section in Chapter 3).

It is vitally important that EMS personnel are effectively coordinated into ongoing operations during technical rescue incidents. Their main functions are to treat patients and to standby in case a rescue team member needs medical assistance. As soon as a technical rescue area or scene is secured, EMS personnel must be allowed access to the victim(s) for medical assessment and stabilization. Some teams have trained paramedics to the technical rescue level so that they can enter hazardous areas and provide direct assistance to the patient. Throughout the course of the operation, which can sometimes span many hours, EMS personnel must continually monitor and ensure the stability of the patient and must be allowed access.

## INCORPORATING "CITIZEN EXPERTS' INTO RESCUE OPERATIONS

Career and volunteer departments may consider recruiting individuals within their communities who have special skills valuable to a technical rescue team, but who may not be interested in becoming a firefighter. Many teams have located search dog handlers who participate in searches but are not required to be trained in firefighting, EMS, or complicated rescue skills. Some teams also include civil engineers, doctors, surgeons, and construction experts.

The inclusion of non-fire service experts in a team is not always a simple matter. These outside members may want the rescue organization to provide injury or malpractice insurance for them. The rescue agency may be concerned about the liability of using outsiders. It has to consider whether it is willing to take on the liability for non-fire service team members during training, during travel to the incident, and at the incident.

An all volunteer department may allow outside experts to join under a special membership category (i.e., special operational members, auxiliary members). In some cases, these personnel may agree to serve on the team at their own risk by signing a waiver releasing the department from having to insure them.

Career departments with no volunteer entity may consider forming a special volunteer organization for the experts to join. The department would have to recognize this organization as an adjunct emergency response entity. This process requires the establishment of corporation bylaws and the application for incorporation to your state. Another option is to pay each expert member one dollar per year, making them employees of the jurisdiction and eligible to receive insurance benefits

from the jurisdictions. Either process usually has to be approved by your city attorney and risk manager.

## MINIMUM NUMBER OF PERSONNEL NECESSARY FOR EACH RESCUE DISCIPLINE

The size of the cadre of personnel comprising the technical rescue team should be based on the type of team and rescue disciplines undertaken, the minimum number of personnel needed to accomplish a rescue mission safely, and the size of the command structure.

Each technical rescue discipline requires its own level of staffing of specially trained rescue personnel. Structural collapse operations, for example, may involve the initial deployment of one or more reconnaissance teams to assess a collapsed structure prior to rescue operations. In general, each reconnaissance team should be comprised of at least three personnel - two specialists working in tandem overseen by a supervisor assessing safety issues. Trench rescue operations are physically demanding and require the movement and construction of heavy panels, timber, mechanical shoring and other specialized equipment. Fewer specialists may be required if a team has advanced, less labor-intensive equipment. Advanced rope operations can be very complex. The more specialty personnel available to simultaneously set up the different parts of a rope system (i.e., raising systems, belay lines, anchoring systems, etc.), the quicker the incident response will be conducted. The desirable minimal staffing level for a confined space entry is two entry personnel backed up by two standby rescuers.

The level of staffing should also be predicated on the number of personnel required to staff command positions (in accordance with established incident management SOPs) as well as the number required to safely and effectively conduct the operation that is undertaken. Other than the normal complement of Incident Command positions (i.e., Incident Commander, Sector Officers, etc.), the technical rescue team should have its own subset of supervisory officers as described in Chapter 9. This may be as simple as four individuals such as Technical Rescue Team Leader,

Technical Rescue Safety Officer, Technical Rescue Equipment Specialist, and Technical Rescue Personnel Specialist.

You should also consider the number personnel which will be committed at an incident and how long they can operate before needing a rest break. If the incident were to last an extended period, do you have sufficient staffing levels for normal day to day operations in addition to the staffing needs of a special incident?

To avoid the unnecessary callbacks of specialty personnel, some departments initially dispatch only one or a few specialty personnel to evaluate the level of response needed. The Tidewater Technical Rescue Team in Virginia Beach, Virginia does not mandate minimum staffing for specialty units, but instead dispatches on-duty personnel as available. Once on scene, the Incident Commander can call for the appropriate number of specialty personnel.

It is important that you specify in the team's operating procedures the minimal number of specially trained and support personnel needed to respond on technical rescue calls or to perform specific functions. Remember that safety at technical rescue incidents is paramount; therefore, if you do not have sufficient trained and qualified personnel to safely execute operations, you should wait until more personnel arrive.

## COMBINATION CAREER/VOLUNTEER STAFFING

Many successful teams use a combination of career and volunteer personnel. Combination staffing for technical rescue teams can combine the benefits of both types of systems. In the area of incident staffing, volunteers can be utilized both on primary teams as full team members or non-team volunteers can be used as supplemental staff to either relieve or fill in for career forces.

The keys to success for combined teams are training and involvement. Both career and volunteer staff have to have the same training levels and train together so that there is mutual respect of each other's technical skills. Similarly, both volunteer and career team members must be included in the rescue team's incident responses. If either is excluded or if one is utilized more

than the other, resentment will build and participation will rapidly drop off.

Including volunteer technical rescue team members on your team offers the potential advantage of volunteer equipment or monetary resources. Volunteers usually have access to resources which can provide equipment vital to your team.

Some departments may choose to staff a technical rescue team with career and volunteer personnel. Both the Fairfax County, Virginia and Montgomery County, Maryland teams benefit from the inclusion of both career and volunteer personnel. The two jurisdictions have found it beneficial to be able to draw from a wide cadre of available personnel who are interested in being involved in technical rescue operations. The addition of volunteer personnel, who have full time careers and offer broader skills/capabilities, have been great assets to their programs. Volunteer personnel with expertise in structural engineering, construction, computers, and emergency medicine are particular assets.

Additional management issues are involved due to volunteer participation. A department must decide if the same training requirements will be in effect for both career and volunteer personnel. Some departments maintain a single standard for both. The major difficulty is ensuring training opportunities for the volunteers. This has been accomplished by ensuring that at least one, if not two, of the three full team training sessions conducted for the three shifts each month is scheduled on the weekend or during the evening. This scheduling allows the volunteers to more readily attend.

## MONETARY COMPENSATION FOR TECHNICAL RESCUE TEAM PERSONNEL

At inception, the motivation and interest of the personnel comprising a new technical rescue team members is very high and provides the momentum that is required to accomplish the wide variety of tasks involved in getting started. This high level of commitment and motivation is difficult to maintain after several years transpire. At some point, the issue of additional or specialty pay for the personnel maintaining technical rescue skills, attending additional training and assuming added risk is sure to arise. It would be in the best interest of the program managers to assess this issue early in the development of a new team. It may be an issue that need not be addressed for several years, but it would be best to have a plan for dealing with this eventuality.

This issue is funding-driven, and developing the justification for additional funding to address specialty pay is paramount. It may be difficult to quantify the level of additional risk inherent in conducting technical rescue operations (although it is certainly present) over and above traditional fire and rescue operations. It may be more prudent to base the justification on the additional skills requirements and training that is required for technical rescue team personnel. These are readily quantifiable.

This issue has been addressed by many technical rescue teams in the nation. Some departments have incorporated the position of Technical Rescue Technician as a position classification within the department rank structure. This is a promotable position requiring the successful completion of a promotional test and ranking on an eligibility list. Promotion to this position

results in a pay increase for the technicians. Dedicated positions at specialty technical rescue stations must be budgeted and staffed.

A somewhat different avenue to the same end may be to request a stipend for specialty skills, also termed proficiency pay. A five to 10 percent stipend in certain jurisdictions is easier to implement and may be somewhat less expensive in the long run (a stipend or specialty pay often does not affect or accrue towards additional retirement pay and benefits as a true promotion might).

Departments which face a difficult time recruiting volunteer members for a technical rescue team due to the time commitment involved may consider paying the personnel on an hourly basis for responses and training. In several Northeast states, the "paid-on-call" concept is used heavily to recruit and retain fire department members. The volunteer department may have to convince local government officials that a small stipend for technical rescue team volunteer personnel is needed due to the time commitment involved, while emphasizing that the services offered by these volunteers are still a bargain.

# CHAPTER 6: REGULATIONS AND STANDARDS GOVERNING TECHNICAL RESCUE OPERATIONS

**T**he formation of any technical rescue team must take into account the various Federal and state laws and regulations that may apply to fire and rescue operations. The most significant regulations are those issued by Federal and state occupational safety and health agencies, which require employers to comply with mandatory minimum workplace health and safety protections. These regulations are based on laws that establish the responsibility of an employer to provide a place of employment that is free from recognized hazards. Violations of Federal and state OSHA regulations can expose a technical rescue team to civil fines and, in rare cases, criminal liability.

The OSHA regulations, however, are not the only legal requirements that a technical rescue team should consider. A number of non-governmental organizations have issued voluntary consensus standards that are relevant to fire and rescue operations. Such organizations include the National Fire Protection Association (NFPA), the American National Standards Institute (ANSI), and the American Society for Testing and Materials (ASTM). Training materials developed by organizations such as the International Fire Service Training Association (IFSTA) and the National Association for Search and Rescue (NASAR) may also be regarded as establishing a "standard" that can be used to evaluate the training and performance of a technical rescue team. Since these standards are issued by private sector associations, they are not binding on fire and rescue departments; however, Federal and state authorities frequently incorporate these standards into their regulations by reference, which may make them legally binding on fire and rescue departments under some circumstances.

A technical rescue team must also consider the impact of the Federal and state laws and regulations, as well as the "voluntary" standards on private litigation. In some states, a technical rescue team may be liable for the negligent performance of their duties. Even in states that protect rescue workers under an immunity statute, most state laws do not protect fire or rescue departments for grossly negligent acts. Essentially,

negligence involves the violation of a standard of care that results in injury or loss to some other individual or organization. In establishing the standard of care for rescue operations, the courts will frequently look to the "voluntary" standards issued by NFPA and other organizations. Although "voluntary" in name, these standards can become, in effect, the legally enforceable standard of care for technical rescue teams. A "specialty team" such as a technical rescue team would be expected to have a higher level of skill and expertise than other individuals, even other members of the same fire department. Accordingly, technical rescue teams should pay close attention to applicable standards of national fire and rescue organizations.

## APPLICATION OF OSHA STANDARDS

The regulations issued by the Occupational Safety and Health Administration of the U.S. Department of Labor (Federal OSHA) are binding *only* upon private sector employers. Therefore, a technical rescue team that is part of a private company must comply with all of the applicable OSHA regulations. An example of a private sector rescue team would include an on-site industrial fire brigade that performs confined space rescue. Such a team would be required to comply with the OSHA Permit-Required Confined Space standard, 29 CFR § 1910.146. In addition, all Federal government agencies, including Federal fire departments, must comply with all OSHA regulations (or issue standards of their own that are at least as effective as the OSHA standards in protecting workers).

State and local government agencies are not subject to the regulations established by Federal OSHA. Although the Federal OSHA standards are not directly applicable to state and local fire departments or rescue agencies, the Federal OSHA law gives each state the option to operate under its own occupational health and safety programs. The states that choose to operate their own program are sometimes called "state OSHA jurisdictions," or simply "OSHA states" and are required by Federal law to cover state and local government employees in the same manner as they do private sector employees. As of July 1995, there were 23 states and two territories which were state OSHA jurisdictions (see Figure 6-1). Although a handful

*Figure 6-1. OSHA States and Territories*

**OSHA States**

| | | |
|---|---|---|
| Alaska | Michigan | Tennessee |
| Arizona | Minnesota | Utah |
| California | Nevada | Vermont |
| Connecticut | New Mexico | Virginia |
| Hawaii | New York | Washington |
| Indiana | North Carolina | Wyoming |
| Iowa | Oregon | (Puerto Rico & |
| Kentucky | South Carolina | Virgin Islands) |
| Maryland | | |

of state OSHA jurisdictions have issued their own regulations that apply specifically to fire and rescue departments (including California, Michigan, and Washington), most state OSHA jurisdictions have simply adopted the Federal standards. This means that in most state OSHA jurisdictions, public sector agencies, including state and local fire departments and rescue agencies, must comply with the Federal OSHA standards. Rescue teams in state OSHA jurisdictions should contact their state occupational safety and health agency to determine what standards apply to them.

## Are Volunteers Exempt From OSHA Laws?

The applicability of OSHA regulations to volunteer fire departments or volunteer firefighters is generally determined by state governments. Generally, under the Federal rules, the applicability of OSHA laws is dependent on the existence of an employer/employee relationship, unless the state has chosen to extend the OSHA laws to volunteers. An employer/employee relationship exists if there is monetary compensation given by the employer to the employees. Federal OSHA does not regulate purely volunteer

operations because there is no employer/employee relationship; however, each state may choose to extend the OSHA regulations or other laws to cover volunteer firefighters. (This is the case in several states including Michigan and New York.) States which are non-OSHA states may have their own regulations over volunteer firefighters, but generally OSHA laws do not apply to volunteers in non-OSHA states.

Generally speaking, if a volunteer receives any type of monetary compensation for responding on a call, attending training, or any other activity, the volunteer can be considered an employee. In some cases where a fire department offers workers compensation protection to volunteers, courts have found that a volunteer can be considered an employee. A volunteer who is reimbursed only for expenses related to fire department operations is usually not considered an employee. Expenses could cover the cost of response in a vehicle to a call.

*Figure 6-2. Application of OSHA Regulations to Volunteer Fire/Rescue Personnel*

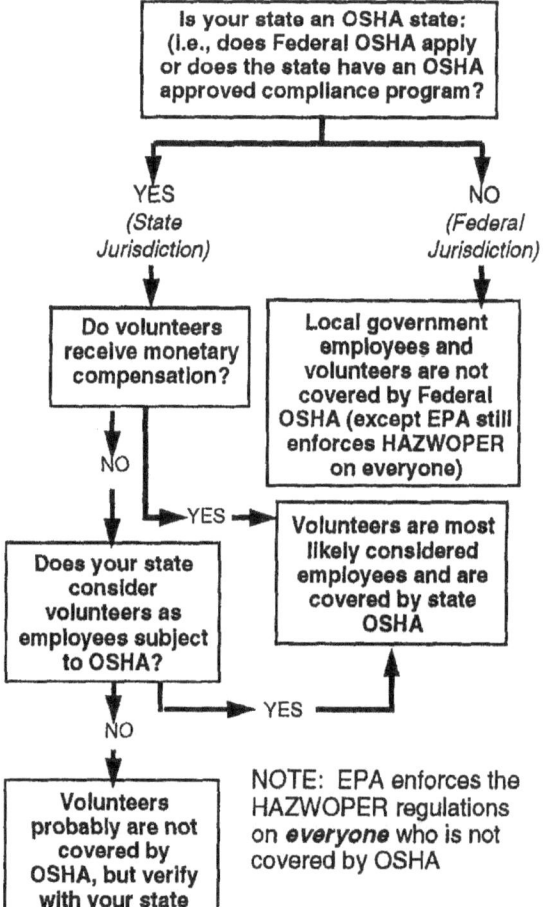

It is wise to talk with your attorney and with your state OSHA office before starting a rescue team. Figure 6-2 will help you determine whether OSHA laws apply to volunteers in your state.

*Regardless of whether or not the OSHA uegulations are legally enforced on your department, they provide important guidance on appropriate safety measures and considerations for many types of situations. Every rescuer should be properly trained and equipped, should have knowledge of rescue hazards, and should know how to perform rescue operations safely.*

## LEGAL CONSIDERATIONS WHEN A PUBLIC TEAM PROVIDES RESCUE SERVICES FOR A PRIVATE ENTITY

Some public sector rescue teams function as the primary response team for industries or as a back-up response team for industrial rescue teams. The industry may be required to provide a rescue capability for its employees who are performing hazardous duties; however, the industry may choose to have the local fire department respond as the rescue team in lieu of forming its own in-house team. In this manner, the fire department may be the first responder, back-up, or support team for an industry.

In OSHA states, public sector rescue teams must comply with their state OSHA regulations as is discussed previously In non-OSHA states, a public sector rescue team can function as the primary or back-up emergency response teams for an industry but the public sector team is not necessarily required to comply with the OSHA standards. The industry, however, is required to ensure that a team responding from an outside agency is capable of meeting the requirements to provide confined space rescue. The industry must also provide access to the potential rescue area(s) for training and familiarization. An industry cannot designate a local fire department as the provider of its required confined space rescue service if that fire department does not have the personnel, training, and equipment necessary to conduct a confined space rescue operation.

## OSHA REGULATIONS AFFECTING TECHNICAL RESCUE TEAMS

As discussed above, state and local government rescue teams in state OSHA jurisdictions are required to comply with all applicable OSHA standards and even volunteer teams may be covered in some states. In non-OSHA states (i.e. a state which does not have its own state OSHA program), even though OSHA regulations may not apply to state or local fire or rescue agencies, technical rescue teams should make every effort to comply with OSHA standards since they can be effective in protecting the health and safety of rescuers.

Federal OSHA regulations are set forth in title 29 of the Code of Federal Regulations. Although most of the regulations can be found in 1901 and 1910, technical rescue teams should also look at 1926, which includes standards for trenching and shoring in the construction industry. An overview of some of the OSHA standards applicable to technical rescue teams follows. Refer to Appendix C for information on how to obtain copies of OSHA standards.

## USC 654(a) ( 1): OSHA General Duty Clause

*Overview.* The General Duty Clause is an all-encompassing section of Federal legislation which describes the responsibilities of employers. The clause is found in 29 USC 654(a)( 1) and states, *"Each employer shall furnish to each of his employees a place of employment which is free from recognized hazards that are causing or likely to cause death or serious physical harm to his employees."* The intent of the clause is to protect employees from workplace hazards establishing a responsibility of the employer to recognize and correct hazards. In general, an employer can be found to have violated this clause for failure to keep the workplace free of a recognized hazard which could have caused death or serious physical harm to an employee and which could have been corrected by some feasible method. Where there is no specific OSHA regulation that applies to a situation, OSHA may use national consensus standards (such as NFPA standards) to determine whether a workplace hazard violated the General Duty Clause.

*Application to Rescuers.* The broad nature of this clause makes it applicable to nearly any fire or rescue related operation where there are hazards present. Basically, it requires rescue agencies to identify hazards to which rescuers might be exposed and to reduce the likelihood of a hazard producing harm to rescuers. This can be done by simply removing the hazard or by providing rescuers the necessary procedures, training, and equipment to safely operate around the hazard.

## 29 CFR 1910.14&
## Permit-Required Confined Spaces

*Overview.* The intent of this standard is to protect personnel who enter permit-required confined spaces. A confined space is defined as an area that:

*1. Is large enough and so configured that an employee can bodily enter and perform assigned work; and*

*2. Has limited or restricted means for entry or exit (for example, tanks, vessels, silos, storage bins, vaults, etc.); and*

*3. Is not designed for continuous employee occupancy.*

A confined space is considered a "permit-required space" if it has one or more of the following characteristics:

*1. Contains or has a potential to contain a hazardous atmosphere;*

*2. Contains a material that has the potential for engulfing an entrant;*

*3. Has an internal configuration such that an entrant could be trapped or asphyxiated by inwardly converging walls or by a floor which slopes downward and tapers to a smaller cross section; or*

*4. Contains any other recognized serious safety or health hazard.*

The term "permit" is used because the standard requires an employer to issue a written permit to employees before they are allowed to enter a permit-required space. The standard goes on to define a hazardous atmosphere as an atmosphere that exposes the employee to the risks of death, incapacitation, impairment of ability to self-rescue, injury, or acute illness from:

*1. A flammable gas, vapor, or mist in excess of 10 percent of its lower flammable limit (LFL);*

*2. Airborne combustible dust at a concentration that exceeds its LFL;*

*3. Atmospheric oxygen concentrations below 19.5 percent or above 23.5 percent;*

*4. Atmospheric concentrations of particular substances with special exposure hazards such as carbon monoxide and hydrogen sulfide; and*

*5. Any atmospheric condition recognized as immediately dangerous to life or health (IDLH).*

This regulation establishes many requirements for employers who have permit-required confined spaces on their premises if they have reason to assign employees to enter these spaces. The portion of this regulation which is most applicable to rescuers, however, is the paragraph K, *Rescue and Emergency Services.*

*Application to Rescuers.* Under paragraph K, *Rescue and Emergency Services,* a rescue agency is required to provide rescuers the full personal protective equipment (including breathing apparatus) and rescue equipment to perform rescues from permit spaces. The rescuers must also be given training on how to use this equipment; no minimum training hours are specified, however, the training must ensure that rescuers are proficient in their assigned duties. Each member of the rescue team must practice making a permit-required space entry at least once every 12 months in a space that is representative of one which they may have to enter. Each rescuer must also be trained in basic first aid and CPR with at least one member holding current certifications in each of these. Paragraph K does not require rescuers to complete a permit before entry is made into a confined space for rescue purposes, however a permit would be required to enter the space for training purposes.

Additionally, under paragraph K and other sections, the regulation establishes that rescuers must have atmospheric monitoring and ventilation equipment, lifelines and harnesses, a mechanical hoist system, communications equipment, and lighting equipment.

## 29 CFR 91910.147:
## Lock-out/Tag-out Requirements

*Overview.* The intent of this standard is to prevent the unexpected energization or start up of machines or equipment, or the release of stored energy which could cause injury to employees. This standard sets forth requirements for the control of hazardous energy or the unexpected start up of equipment.

*Application to Rescuers.* Lock-out/tag-out procedures may be necessary when performing rescues involving heavy industrial equipment, elevators, or electrical rooms. Electricity must be shut down and protected so that re-energizing does not occur while the rescue is being performed. Employers must create an employee protection program that defines Lock-out and Tag-out procedures.

## 29 CFR §1910.132-.140:
## Personal Protective Equipment

*Overview.* This section establishes general requirements for the employer to provide, test, inspect, and maintain personal protective equipment (WE) for employees who are exposed to workplace hazards. Employees must be trained on proper use of equipment. 1910.134 addresses respiratory protection and requires that where employees must enter a hazardous area using PPE, one or more employees equipped with PPE must be assigned to stand by to provide accountability and to assist in the rescue of entrants in the event they need emergency assistance. The backup requirements depend on the nature of the hazard.

*Application to Rescuers.* This standard requires rescue agencies to provide rescuers with the necessary personal protective equipment to safely enter a hazardous environment, including eye protection, face protection, head and extremity protection, protective clothing, respiratory protection, and protective shields and barriers. Rescuers must be trained on the proper use of the equipment and the rescue agency must have written operating procedures for its safe use. Specific requirements are listed for regular maintenance and testing of respiratory equipment, fit testing, and other requirements for a respiratory protection program. The requirements for backup personnel also apply to rescue operations.

All of the basic requirements that are applicable to employees in general industries also apply to firefighters and rescue workers. There are additional requirements that apply specifically to fire suppression and rescue operations.

## 29 CFR §1910.1030:
## Occupational Exposure to Bloodborne Pathogens

*Overview.* The intent of this section is to provide for employee protection from exposure to bloodborne pathogens or other potentially infectious materials.

*Application to Rescuers.* This standard requires that rescue agencies provide a comprehensive education and control program for rescuers who may be exposed to bloodborne pathogens or infectious materials. The program must cover the following topics: training for rescuers about the dangers of bloodborne pathogens; how to dispose of contaminated materials; disposal processes for sharps, contaminated instruments, and infectious materials; documentation of rescuer exposures to infectious materials; and post-exposure medical evaluations. The rescue agency is required to provide all protective equipment that is necessary to protect the employees from bloodborne pathogens. Hepatitis B vaccinations must be offered at no cost to rescuers.

## 29 CFR §1926.650 - §652:
## Trench/Collapse Rescue Operations

*Overview.* This section establishes operational and safety practices for incidents involving trenches.

*Application to Rescuers.* This regulation directly affects rescue operations at trench rescue incidents by requiring the use of proper equipment and techniques to shore-up trenches. It prohibits entry into trenches which are not properly shored to prevent collapse. This standard also specifies that rescuers wear a lifeline into trenches. It also requires that rescue agencies provide training to rescuers about the hazards of trench operations.

## OTHER OSHA REGULATIONS

Some of the other OSHA regulations which can affect technical rescue teams are listed below and discussed in Figure 6-3.

*29 CFR §1910.95: Occupational Noise Exposure Limitations*

*29 CFR §1910.156: Fire Brigades*
*29 CFR §1910.120: Hazardous Materials*

## CONSENSUS STANDARDS AFFECTING TECHNICAL RESCUE TEAMS

Consensus standards are developed by industries and other groups to set forth widely accepted criteria for care and operations for certain practices. Standards are an attempt by the industry or profession to self-regulate by establishing minimum operating, performance, or safety standards, and they establish a recognized standard of care. They are written by consensus committees composed of industry representatives and other affected parties. Several of the NFPA standards that are pertinent to technical rescue teams are discussed below.

When forming a rescue team, the standards issued by national fire and rescue organizations should be followed to protect rescuers from

*Figure 6-3. Selected OSHA Standards Applicable to Technical Rescue Teams*

| Standard | Basic Requirement | Equipment | Training | Miscellaneous |
|---|---|---|---|---|
| General duty clause 29 USC 654(a)(l) | Requires employer to provide a workplace "free from recognized hazards that are causing or likely to cause death or serious physical harm." OSHA frequently uses "national consensus" standards (i.e., NFPA standards) in finding that a workplace hazard violated the GDC. | Although the GDC does not list specific equipment, courts have ruled that employers must provide whatever equipment is needed to protect workers and eliminate unreasonable risks of death or injury. Employers can be cited under the GDC for not providing necessary equipment or for providing equipment that does not meet nationally recognized standards (i.e., NFPA standards). | As with equipment, the GDC requires employers to provide whatever training is necessary to allow employees to perform their assignments in a reasonably safe manner. | Employees may be cited under the GDC where (1) employer failed to keep workplace free from hazard; (2) hazard was recognized; (3) hazard was likely to cause death or serious injury; and (4) a practical method for correcting the hazard was available |
| Confined space 29CFR 1910.146 | Classifies confined spaces as "permit" or "non-permit" spaces. Requires permit before entry into spaces with potential hazards. Sets basic safety practices for work in confined spaces. Separate provisions apply to rescue operations. | Employers must provide full PPE, including SCBA or SABA. Rescuers must have atmospheric monitoring and ventilation equipment, harnesses and life lines, and mechanical hoist systems, Lockout/tagout, communications, and lighting equipment should be available. | Rescuers must receive training sufficient to perform rescues; no minimum number of hours is specified. Rescuers must perform practice entries and rescues from realistic spaces at least annually. Rescuers must have CPR and basic first aid training, and one member must be currently certified in both. | Requires companies that rely on outside rescue services to provide access to confined spaces for pre-planning and training purposes. Standard includes recommended entry permit and flow chart for rescue operations |
| Blood-borne pathogens 29CFR 1910.1030 | Requires employers to develop written plan to reduce exposure to blood-borne pathogens (BBP). | Employers must provide PPE that does not permit blood or other infectious materials to penetrate to employee's skin or clothing. Standard contains a detailed listing of such equipment. | Employers must provide training on the hazards of BBP and on procedures to prevent exposures. Training courses must contain 14 elements, including written materials and an opportunity to ask questions. Refresher training must be provided annually. | Employers must provide hepatitis B vaccine at no cost to employees. Training records must be maintained for at least three years, |
| Personal protective equipment (PPE) 29 CFR 1910.132-140 | Establishes minimum protective equipment that employers must provide. | Minimum PPE includes eye, face, head, hand, foot, and leg protection, as well as structural fire fighting gear. Equipment must meet ANSI, NFPA or other industry standards, | Employees must receive training in proper use and maintenance of all equipment. | Requires written procedure for use of respirators. Requires employees working in toxic atmospheres to work in pairs and at least one person capable of rescue must remain outside toxic atmosphere. |

unnecessary workplace hazards and because they establish the standard of care that may be used in civil lawsuits against fire and rescue. Several standards related to technical rescue are discussed below. Refer to Appendix C for information on how to obtain copies of standards.

*Figure 6-3. Selected OSHA Standards Applicable to Technical Rescue Teams (Continued)*

| Standard | Basic Requirement | Equipment | Training | Miscellaneous |
|---|---|---|---|---|
| Trench/collapse rescue 29 CFR 1926.650-652 | Establishes safety guidelines for construction work Involving open trenches or excavations. Separate provisions apply to rescue operations. | Employers must provide shoring, harnesses and lifelines, and basket stretcher. Type and amount of shoring equipment varies according to trench to be entered. | Employees must receive training In hazards of trench and excavation activities, Including proper use of shoring and other equipment. | Applied to fire departments, prohibits fire fighters from entering unshored trenches or taking other unreasonable risks. |
| Hazardous materials 29 CFR 1910.120 | Requires employers to develop comprehensive program for "hazmat" teams and establishes minimum safety standards. 1910.120 applies to all personnel Involved In hazmat response, volunteer and paid. in every state. | Employers must provide full PPE for hazmat Incidents. Appendix A of this standard offers guidelines for testing such equipment, while Appendix B discusses different levels of protection and gear. | Establishes five levels of training certification: awareness, operations, technician, specialist, and Incident command. All levels must receive annual refresher training. Trainers must have completed National Fire Academy or similar course. | Employers must develop comprehensive written hazmat response plan. The plan must address pre-planning, personnel roles, EMS, decontamination, and emergency alerting. ICS must be used at hazmat Incidents. Employers must Implement medical surveillance program for members of hazmat team. |
| Fire brigade 29CFR 1910.156 | Establishes minimum administrative, Personnel equipment and training standards for fire brigades. In state-OSHA plan jurisdictions, fire brigade standard applies to public sector (state, county, and municipal) fire departments. | Must provide employees with the fire fighting and personal protective equipment necessary to perform job. Protective gear must meet NFPA or other industry standards. | Employees must be trained to level required on the job. Supervisory personnel must receive additional training. Training must be of similar quality to that offered at specified schools across country. | Written organizational statements must specify duties and establish amount and type of training. Employees must be physically capable of performing assigned firefighting duties. |

## NFPA 1470:
## Standard on Search and Rescue Training for Structural Collapse Incidents

*Overview.* This standard sets forth requirements for training, safety, operations and personal protective equipment at structural collapse incidents. The standard was developed following a number of different major structural collapse incidents across the country. NFPA 1470 defines four levels of response operations which depend on an organization's capabilities, equipment, training, certifications:

1. **Basic Operations,** which covers surface rescue at structural collapse incidents including the removal of debris to extricate easily accessible victims in stable environments.

2. **Light Operations,** which is a minimum capacity to conduct a safe and effective search and rescue where the collapse is of a light-frame ordinary construction building.

3. **Medium Operations,** which covers a response to a building or structural collapse involving the failure of cinder block and non-reinforced masonry construction.

4. **Heavy Operations,** which covers structural collapse involving the failure of concrete tilt-up or reinforced concrete and steel frame construction.

*Application to Rescuers.* NFPA 1470 is the only NFPA standard that directly addresses technical rescue. It sets forth minimum training, personal protective equipment, and operational standards for rescuers to develop systems for structural collapse incidents. The standard mandates use of an incident management system at structural collapse incidents and requires oversight by a qualified safety officer. All persons must be physically and mentally capable of performing the appropriate duties and functions for each level of response.

NFPA 1470 mandates training for recognition of various hazards at collapse incidents including but not limited to hazardous materials. It also requires that certain rescue team members be trained to treat various medical conditions including spinal injuries, crush injury syndrome, and amputation. Provisions must be made for infectious control (including education on OSHA's bloodborne pathogens that is incorporated by reference) and critical incident stress debriefing.

NFPA 1470 is currently being expanded by a new NFPA technical committee to cover a broader range of technical rescue disciplines, not just structural collapse. The new, expanded standard is expected to cover confined space rescue, water rescue, structural collapse rescue, trench rescue, wilderness rescue, and vehicle/machinery rescue (although not officially determined yet, NFPA 1470 may be incorporated into the new standard). It is proposed to create three rescue response levels: awareness, operations, and technician. The new standard will shape technical rescue in the future and will probably serve to standardize technical rescue training across the country. The new technical rescue standard will probably be issued by 1997.

## NFPA 1500: Standard of Firefighter Health and Safety

*Overview.* NFPA 1500 establishes minimum standards for firefighter health, safety, and fire department risk management. It applies to all aspects of the workplace, including the emergency scene and the station.

The standard requires fire departments and rescue agencies to:

- Establish an organizational/mission statement

- Establish a minimum number of members, resources, and equipment required to safely execute standard operating procedures

- Establish risk management plans (fire and rescue pre-incident plans) for the purpose of identifying the risks, evaluating potential risks, and establishing appropriate risk control techniques and risk management monitoring techniques (command and control techniques)

- Adopt SOPs identifying specific goals for the various operations that may be encountered by the department, including rescue operation

- Define roles and responsibilities for responding personnel

- Provide continuing education for fire and rescue personnel, detailing the frequency of training and minimum requirements

- Establish safety requirements for emergency vehicles and equipment

- Utilize incident management, risk management, and personnel accountability systems at incidents

- Establish firefighter medical and physical fitness requirements

- Create a department Safety Officer position and an Occupational Safety and Health Committee

- Establish data collection systems to track occupational exposures, health records, training records, and equipment inspection records

*Application to Rescuers.* NFPA 1500 is a very broad standard with applications to all aspects of fire service response including technical rescue. For a technical rescue team, it requires the establishment of SOPs, utilization of incident command, personnel accountability, and safety systems at incidents, and the determination of the number of personnel necessary to conduct a safe rescue operation. It also requires use of personal protective equipment as necessary for the hazards and requires that responders meet medical/fitness levels.

## NFPA 1983:
## Standard on Fire Service Life Safety Rope, Harnesses, and Hardware

*Overview.* This standard specifies the minimum performance criteria, design criteria, and test methods for life safety rope, harnesses, and hardware used by the fire service. It does not, however, cover rope equipment used for water rescue. It also establishes requirements for inspection and testing of rope and related hardware, as well as minimum working load strengths.

*Application to Rescuers.* This standard establishes the types of rope that should be used for life safety (securing a rescuer, hauling or lowering a rescuer, etc.). It also establishes the amount of weight that can be supported on a rope of particular design, fiber, and diameter. There are requirements for regular inspection and testing of rope by rescuers.

## OTHER STANDARDS

Some of the other NFPA standards affecting technical rescue teams are listed below. Contact the NFPA for copies of these standards (refer to the resource list in Appendix C).

*NFPA 1561: Standard on Fire Department Incident Management Systems*

*NFPA 1001: Standard on Professional Qualification for Firefighters*

*NFPA 1581: Standard on Medical Requirement for Firefighters*

*NFPA 1582: Standard on Fire Department infection Control Program*

# CHAPTER 7: TECHNICAL RESCUE TRAINING

**N**o tools or technology can compensate for lack of training and experience. Proper training is necessary for any rescue team to safely and effectively conduct rescue operations. This chapter discusses the evolution of technical rescue training, the future of rescue training, training requirements, how to plan training for your department, and curriculum for different training levels.

## THE DEVELOPMENT OF TECHNICAL RESCUE TRAINING

To understand how the field of technical rescue is currently developing, a review of the history of similar fire department operations is useful. In the 1960s, many fire and rescue organizations provided a basic level of emergency medical care and transportation following procedures that had changed little in decades. The training levels of emergency medical providers varied considerably from jurisdiction to jurisdiction and from state to state, and there were no established national standards except those basic ones provided by the Red Cross. In the late 1960s, EMS professionals began to band together and form networks to exchange ideas. Common practices evolved into standards of care at the local, state, and eventually national level. First responder, emergency medical technician (EMT), and paramedic training programs were started in many areas and nationally recognized training programs were widely adopted. Today, we are all familiar with the basic EMT curriculum and a consistent level of EMT care is available across the country. Hazardous materials programs followed a similar course of development in the 1980s and a similar system of national training standards for hazardous materials responders was established.

*Figure 7-1. Selected Standards Affecting Technical Rescue Training*

| Rescue Discipline | OSHA Standard | NFPA Standard | Comment |
|---|---|---|---|
| Confined Space | 29 CFR §1910.146 | None | 29 CFR 1910.146 training requirements mandate annual entry training at a representative permit space and basic first aid training, but do not specify levels of training or minimum training proficiencies. A separate OSHA standard on hazmat operations training, 29 CFR 1910.120, affects training for operations in confined space with IDLH (toxic or oxygen deficient) environments |
| Collapse | None | NFPA 1470 | |
| Water/Diving | None | None | Professional Association of Diving Instructors) PADI and other dive organizations have standards for dive training. The American Red Cross also has water rescue training standards |
| Trench | 29 CFR 1926.65 0-.652 | None | 29 CFR 1926.650-.652 mandates training on hazards of trench activities, including proper use of shoring, but does not establish operational training levels |
| Rope | None | None | NFPA 1983 is the standard for rope to be used for rescue but does not discuss training |

Over the years, a variety of organizations and private training companies have established training standards for the different technical rescue disciplines. Some of the organizations which have created rescue training standards include: OSHA, NFPA, the American Society for Testing and Materials (ASTM), the American Red Cross, the National Association of Search and Rescue (NASAR), and a host of private training companies. Figure 7-1 shows some of the standards affecting rescue disciplines. These standards are discussed in detail in Chapter 6. Some of these standards have been recognized and adopted by various states or localities, but there are very few nationally recognized training standards directed at technical rescue. In the absence of national training standards, some of them have been used to establish certification levels at the state and local level.

## THE FUTURE OF TECHNICAL RESCUE TRAINING STANDARDS

Technical rescue is beginning to undergo the same pattern of development that EMS did in the 1970s and hazmat did in the 1980s. Rescue training continues to become more formalized as various training standards are issued. In the future, more training standards may become mandated if adopted by state or Federal OSHA. The NFPA, ASTM, and other organizations and private entities most likely will continue to shape rescue training by issuing their own standards. Both the NFPA and ASTM are working on comprehensive national rescue training standards, but it remains to be seen if these will become the "nationally" accepted standards for rescue training. The NFPA standard is discussed below.

One of the major advantages to standardizing training is that rescue personnel across the country will receive similar training at specific response levels. This allows training organizations to develop standard programs that can be widely used and identifies procedures that are accepted as safe and effective for rescue operations. This should also allow personnel on teams from different jurisdictions to work together and permit greater portability of training certifications from one jurisdiction to another.

## THE NFPA STANDARD ON TECHNICAL RESCUE

In 1994, a NFPA committee began working on the development of a document that will establish national training standards for several technical rescue disciplines, including confined space rescue, water rescue, structural collapse rescue, trench rescue, vehicle/machinery rescue, and wilderness rescue. This standard is not expected to be issued until at least 1997.

The NFPA standard on technical rescue training is expected to structure training around three or four levels of response, similar to the tiered response levels discussed in Chapter 1. These include: basic or awareness, operations, technician, and specialist or instructor.

In the first level of training the awareness level, personnel receive basic training to recognize technical rescue situations and the hazards that may be involved so that they can call for additional resources and do not expose themselves to dangers they are not trained or equipped to handle. At this level, they know what they can and cannot do, when to call for specialized assistance, and what to do until the appropriately trained personnel arrive.

The second level of training is the operations level. Operations level personnel receive additional training in hazard recognition, and are trained to perform moderately complex rescue operations.

The third level of training is the technician level. Technician personnel should be well trained in advanced techniques and capable of conducting more complex technical rescue operations.

A fourth level of training which may be included is the specialist or instructor level. Specialist personnel are highly qualified in one or more aspects of technical rescue operations. Specialists should be experts in the use of various rescue technologies, and thoroughly familiar with all aspects of rescue operations in their particular area(s).

## SOURCES OF TRAINING

There are many sources of rescue training available. There are private companies that will provide training in particular rescue disciplines. Many public agencies and fire departments also offer rescue training, particularly for personnel from public safety organizations. Most of these types of courses will "certify" that the student has completed the course and has achieved a minimum level of competency. However, the competency levels taught by individual trainers often vary due to the lack of standardization in rescue training.

Several states have established their own certification levels and training courses available for different areas of technical rescue. California, for example, has a rescue certification program that is under the State Fire Marshal's Office. Maryland and Virginia also have a state rescue training course. Some local jurisdictions have established their own training curriculum (see Figure 7-2). Montgomery County, Maryland, teaches a "tactical rescue" course which provides basic training to rescuers in trench/collapse, rope, confined space, extrication, and swift water rescue. Indianapolis, Indiana, has established a basic emergency rescue technician program that provides similar training.

Appendix C lists different groups that can be a source of technical rescue training.

## DEVELOPING A TECHNICAL RESCUE TRAINING PLAN

It is important to develop training plans from the initial stages of team development. In many cases, members of departments take training courses on their own and then develop a team on their own out of shared interest and competence in the subject. In other cases, members have no formal training whatsoever and are trained after the team concept is officially formed by their organization.

Several factors will affect the type of training program necessary. These factors are discussed below.

### Your Response Area

A general knowledge of technical rescue can be imparted through training, but one of the most important factors in developing a training program that meets your locality's need is the nature of your response area. Training should be directed toward the geography and target hazards in the team's response area. Technical rescue training techniques can then be adapted to train you for responses to these hazards. Training should incorporate a thorough and systematic overview of the potential technical rescue hazards in the team's response area. The team should develop pre-incident plans for the target hazards and train on rescue scenarios that could occur. Training is not complete without a thorough knowledge of how to handle rescues involving the hazards in the team's response area.

### Type of Team

It will be important to decide whether a multi-discipline or a single discipline team will be necessary. Depending on the type of team, how many personnel will be trained to the awareness/operations level; how many to the technician level; how many to the specialist/instructor level?

### Training Resources Available

Consider the training resources that are available within your department or from outside sources within your community. Are there experts within the organization that can train your personnel, or will it be necessary to bring in outside instructors? Can training be taught at a convenient location such as a fire academy or local river or local climbing area, or will personnel have to be transported to another location? Is additional equipment needed to conduct the training?

### Available Funding

Training can be an expensive undertaking. Is the money available for training an advanced team, or will the jurisdiction have to accept a basic level of response capability? How much money is available for equipment, personnel, etc.? (Refer to Chapter 4.) Remember, the lack of training can be much more expensive.

### Personnel

The availability of personnel also affects training. Will training schedules have to accommodate volunteer personnel who may only be available on nights or weekends? Are personnel part of companies that can be rotated to training drills or in-service programs? Is the team made up of members from different organizations with dissimilar schedules? Will career departments fund overtime or additional staff hours for specialized training?

*Figure 7-2. Fairfax County (VA) Fire and Rescue Department Training Curriculum*

**FAIRFAX COUNTY FIRE AND RESCUE DEPARTMENT**
**TECHNICAL RESCUE OPERATIONS TEAM**

**Certification Training Curriculum**

- **Module One: Advanced Rope Rescue Operations** (three days)
  - Anchor/lowering/rappelling systems
  - Mechanical advantage/raising systems
  - Stokes litter rigging and raising systems
  - Tyrolean systems
- **Module Two: Trench Rescue Operations** (three days)
  - Trench hazards/soil mechanics/contributing conditions
  - Review of case studies
  - Scene management procedures
  - Hazard mitigation/terminology/specialty equipment
  - Shoring procedures (timber, screw jacks, pneumatic)
  - Alternate methods of "safeing" a trench
  - Extrication/patient removal techniques
  - Termination of operations
- **Module Three: Confined Space Entry and Rescue Operations** (two days)
  - Defining the confined space parameters
  - Familiarization with OSHA confined space regulations
  - Operational hazards (atmospheric, physical, psychological)
  - Hazard mitigation/monitoring/ventilation
  - Isolation and lock-out procedures
  - Respiratory protection (SCBA, supplied air systems)
  - Support operations (lighting, communications, etc.)
  - Entry/egress methods
  - Rotation of crews/accountability/health assessment
  - Patient removal techniques
- **Module Four: Structural Collapse Operations** (two days)
  - Overview of collapse
  - Collapse cause classification
  - Strategic objectives
  - Operational plan of action/scene management
  - Reconnaissance team duties
  - Gaining entry/tools considerations
  - Exploration of voids
  - Rescue team/safety team operations
  - Shoring guidelines
  - Estimation of weights of building components
  - Search technology and considerations
  - Researching the building
  - Selected debris removal
  - Consideration for cranes and heavy equipment

*Courtesy of Fairfax County Fire and Rescue Department*

Once you have determined what type of training your department needs, begin to shop around for available training courses. Compare prices, the experience of instructors, and what training competencies will be met in the courses.

It is helpful to have a special operations training coordinator or committee to steer the training process. Personnel from all ranks should be included in the planning. It is a good idea to train some of the team leaders first; this will give them an idea of what will be required to train the team members.

The plan for training should include the number of personnel, the desired level of training for those personnel, recertification requirements, equipment needs, and locations for classroom and field training. It is not necessary to reinvent the training wheel, so to speak, but certain differences between technical rescue training and fire training are important to point out.

When conducting technical rescue field training, it is usually advisable to have a higher ratio of instructors to students than with most fire training, due to the complex and sometimes dangerous nature of the training. Safety officers should be assigned whenever the training involves a potential risk of injury. Swift water rescue training may require additional personnel downstream from the training as safety monitors.

Technical rescue training may be more equipment intensive than other types of training. Teams should be prepared to use and maintain this equipment and to replace it when necessary. Training exercises should use the same equipment that the trainees will actually use for real incidents.

It is a good idea to refer to standards such as NFPA 1403, *Standard on Live Fire Training Evolutions,* and NFPA 1500, *Standard on Occupational Safety and Health,* when actually conducting training. The basic safety approaches and requirements in these standards can be applied to rescue training and will help ensure a safe training environment. These include pre-planning for the drill, using a Safety Officer, having rescue personnel available should the students run into trouble, having EMS personnel and equipment on hand, and using the Incident Command System.

## SPECIFIC TECHNICAL RESCUE TRAINING EXAMPLES

To give departments an idea of the various technical rescue training curriculum that could be established, sample outlines of some types of technical rescue course curriculum follow. These sample outlines are intended only to present some of the topics which could be covered in curriculum and are not necessarily complete outlines.

### Rope Rescue

Rope techniques are a basic underlying skill for most other types of rescue. Most firefighters will be familiar with basic rope techniques and knot tying as part of the basic firefighter curriculum.

An awareness of rope skills can be taught to rescuers in only a day. It could include topics such as rope characteristics, strengths, basic knots, hardware, hazards to be aware of when using rope, and dangerous techniques to avoid.

An operations level could cover rope rescue techniques. Rescuers could be taught basic techniques of rappelling, rigging, belaying, safety, anchoring, and simple mechanical advantage systems. Additional operational techniques could include patient packaging, low angle evacuations, and simple pick-off maneuvers. This could be taught in two days.

A detailed technician level program could be conducted in approximately one week, covering basic and advanced rigging techniques, anchor systems, belays, simple and complex mechanical advantage systems, and advanced patient extrication techniques and stokes basket operations. Low and high angle rescue techniques, including telpher and tyrolean systems, could also be included.

The specialist level course could include advanced techniques for helicopter operations, ladder operations and bridging techniques, and other topics. It should require practical and teaching experience. Urban rope techniques could be incorporated for areas where high angle rescues may be adapted to an urban environment.

*Sample Course Topics*
- Course *objective*
- *History of rope rescue*

- *Rope rescue applications*
- *Rescue philosophy*
- *Safety*
- Types of rope
- *Types* of *equipment*
- *Types* of *hardware and technical gear*
- *Communications*
- *Knots, hitches, and anchors*
- *Lashing and picketing techniques*
- *Simple and complex mechanical advantage systems*
- *Belay techniques*
- *Litter rigging and evacuation techniques*
- *Low angle rescue*
- *High angle rescue*
- *Urban rescue operations*
- *Traverse techniques*
- *Incident command*
- *Selfrescue techniques*
- *EMS and patient care considerations*
- *Helicopter operations*

### Personal Equipment
- *Helmet*
- *Sturdy boots*
- *Leather gloves (preferably not firefighting gloves)*
- *Harness*
- *Clothing (appropriate for terrain and weather conditions)*

## Confined Space Rescue

Confined spaces are defined as any area not designed for human occupancy with limited entrance and egress (refer to the discussion of confined space regulations in Chapter 6 for a full definition). The Occupational Safety and Health Administration has established one of the few standards that is applicable to technical rescue, 29 CFR §1910.146, which requires confined space rescue personnel who enter permit spaces to be trained (although it provides little training specifics, as discussed earlier in this chapter).

An awareness of confined space rescue can be taught in a few hours. The awareness level for confined space could include background on OSHA regulations, recognition of permit-required spaces, confined space hazard recognition, how to secure the scene, available resources for confined

space rescue, and what conditions preclude their entry into a space.

Operations level personnel could be taught safe entry and rescue techniques, atmospheric monitoring techniques, and how to size up the hazards and risks. An operations level could be achieved with several days of training.

Technician level personnel could be trained for a wide range of skills and hazard assessment. Skills may include patient evacuation, special retrieval systems, use of communications and command at confined space incidents, familiarity with various types of confined space, atmospheric monitoring, hazard assessment, and ventilation techniques. At least 40 hours would be necessary to train personnel to the technician level.

The specialist should be fully versed in confined space operations and have hands on, practical experience. A specialist should have the expertise of the technician, along with experience in training, hazardous materials, and other associated rescue areas that would be applicable to confined spaces.

### Sample Course Topics
- *Types of confined spaces*
- *OSHA rules*
- *Hazard recognition*
- *Securing the scene*
- *Resources*
- *Atmospheric monitoring*
- *Incident command*
- *Rescuer entry techniques*
- *Retrieval systems*
- *Rope and hardware and technical equipment*
- *Lock out/Tag out procedures*
- *Breathing apparatus equipment*
- *EMS and patient care considerations*
- *Safety and survival*

### Personal Equipment Necessary
- *Helmet*
- *Gloves*
- *Work boots*
- *Personal protective clothing*
- *Harness*
- *Knee pads/elbow pads*
- *Eye protection*
- *SCBA/supplied air breathing system*

## Trench Rescue

By definition, a trench is deeper than it is wide. Rescuers have been killed and injured after entering an unshored trench which suffered a secondary collapse. Awareness of the dangers of trench incidents can be taught in about two hours, covering the basics of hazard recognition, scene security, rescuer safety, types of trench collapses, additional resources, and initial actions.

An operations level of training can be taught in several days, with students gaining knowledge of rescue equipment, different types of shoring, means of securing the site according to departmental SOPs, how to perform a safe entry, and other support operations.

Technician level personnel could become familiar with various rescue techniques, shoring techniques, victim retrieval systems, EMS and patient care skills for trench collapse, control of utilities, and long term operations skills. The technician level could be taught in about 10 days.

A specialist could be thoroughly expert in the use of all types of rescue equipment and techniques for trench rescue incidents and should have practical and teaching experience.

Trench rescue shares equipment, rescue techniques, and skills with both confined space rescue and collapse rescue. A course could be designed to include aspects of each discipline.

### Sample Course Topics
* Trench hazards
* Securing the scene
* Safety
* incident command
* Equipment and resources
* Department SOPs
* Shoring techniques
* Rigging
* EMS care
* Entry and patient removal techniques

### Personal Equipment
* Helmet
* Gloves
* Work boots
* Personal protective clothing
* Harness
* Knee pads/elbow pads
* Eye protection
* SCBA/supplied air breathing system
* Folding shovel

## Structural Collapse

Structural collapse shares many techniques with trench and confined space rescue. An awareness of the dangers of structural collapse could cover types of construction and associated hazards, types of collapses, how to secure the scene, and when to call for help. This could be taught in approximately eight hours.

An operations level of training could also include patterns for conducting a surface debris search for victims, basic stabilization, utility control, and atmospheric monitoring. It could be taught in two to three days.

A technician level course covering shoring and building stabilization, rescue equipment, search equipment and operations, tunneling and excavation techniques, and patient care could be taught in approximately five days.

A specialist should be expert in the use of various types of light and heavy rescue technologies, hazard stabilization and mitigation, and the components of urban search and rescue techniques.

### Sample Course Topics
* Size up and command considerations
* Construction types
* Types of collapses
* Initial actions
* Dangers to rescuers
* Basic search techniques
* Advanced search techniques
* Shoring and stabilizing techniques
* Equipment and technologies for collapse rescue
* EMS and patient considerations
* Safety and psychological impact/critical incident stress debriefing
* Breaching concrete and steel and other barriers
* Tunneling and excavation techniques
* Hazards to rescuers
* Heavy construction equipment operations

### Personal Equipment
* Helmet
* Gloves

- *Work boots*
- *Personal protective clothing*
- *Harness*
- *Knee pads/elbow pads*
- *Eye protection*
- *SCBA/supplied air breathing system*
- *Folding shovel*

## Water Rescue

One of the most dangerous types of special rescue is water rescue. There are several different specialties within the field of water rescue. Rescuers may face incidents involving calm water, swiftwater, ice, or even surf conditions. Dive rescue is a specialty within itself. Courses in each training level could be designed to address all types of water rescue or individual types (e.g. swiftwater rescue only).

A basic awareness of water hazards, safety, and shore-based rescue techniques can be taught in a few hours. Different types of water rescue may share similar techniques, but pose different dangers.

Operations level training could cover techniques for in-water or ice rescue. Rescuers could become familiar with different types of water rescue techniques, ice and current hazards, hypothermia and EMS considerations, ice rescue equipment, and shore-based swift water rescue techniques. This course could be taught in about one week, but would require personnel to be able to swim.

The technician level could require knowledge in all facets of water rescue and how to perform special rescue techniques such as victim retrieval using boats or a helicopter. This course could be taught in about one week.

The specialist level could require in-depth knowledge of all types of water rescue techniques and hazards as well as practical and training experience.

### Sample Course Topics

- *Water hazards*
- *Ice characteristics an dangers*
- *Swift water hazards and hydraulic charcteristics*
- *Reach techniques*
- *Throw techniques*
- *Row techniques*
- *Go techniques*

- *Helicopter uses*
- *Cold water drowning and hypothermia*
- *Self rescue and survival techniques*
- *Rescue vs. recovery*
- *Diver support*
- *Search patterns and techniques*
- *Safety*
- *Incident command*
- *Boat operations*
- *Flashflood and rising water*
- *Contaminated bodies of water*
- *Ice rescue equipment and techniques*
- *Swift water rescue equipment and techniques*
- *Surf rescue equipment and techniques*
- *Basic water safety*
- *Swimming test*

### Personal Equipment

- *Personal floatation device/life vest*
- *Whistle*
- *Knife or shears*
- *Flashlight*
- *Rope throwbag*
- *Helmet*
- *Gloves*
- *Goggles/eye protection*
- *Wet or dry suit*
- *Suitable footwear*
- *SCUBA gear (dive team only)*

## RECERTIFICATION AND CONTINUING EDUCATION

Recertification for technical rescue personnel is necessary to refresh practical skills and knowledge about the subject matter. In all types of technical rescue, skills must be honed and practiced to maintain a high readiness level. New technologies and new techniques are constantly being developed to make technical rescue operations easier and safer. It is important to allow for continued training beyond basic training. Teams will learn to work together better, and an exchange of ideas and information will allow knowledge to be spread among experienced rescuers. Unlike EMS, a set amount of hours may not be necessary, but an annual, skill based test in competency, with the ability to retrain in deficient areas, may be the best way to keep an individual's skills and a team's level of competence consistent.

## DOCUMENTATION

Documentation should be kept for individuals, the team, and equipment, for both training and actual incidents.

### Individual Records

Teams should keep records of all training, including initial training and certification, and continuing education training for each student. Documentation should include training hours, skills demonstrated, skills performed, and skills tested. Evaluations by instructors and supervisors should be included.

### Team Records

Documentation should also be kept for the team as a whole, including types of training, hours, equipment used, and costs incurred. Use of new equipment and techniques, along with their limitations and advantages, should also be recorded. Personnel should be tracked for their level of training, readiness, and injuries.

### Equipment

A log of major equipment, including life safety equipment such as breathing apparatus or rescue rope, should be kept to track use, repairs, problems, and replacement. This will help maintain a record should questions arise about a piece of equipment's use or safety.

### Incident Records

It is vital to conduct a thorough review of each technical rescue incident and to document it. This will allow teams to understand what occurred and to develop strategies to improve the safety, efficiency, and effectiveness of their training and preparation for future incidents.

Record keeping serves two main functions. First, it allows a team to establish a baseline for their readiness capability, so that they may use performance based criteria to improve their operations. It also allows them to chart their progress and discover during periodic review the areas that need improvement. Secondly, record keeping provides much needed documentation should legal issues arise from team operations.

## TEAMWORK

One of the most important aspects of training in technical rescue is to teach rescuers to function as a team. Difficulties can arise when individuals do what they think is best, often working alone, inefficiently, and dangerously. Problems can also arise if rescuers from different companies or different organizations are forced to work together without having previously trained together. These problems can be overcome by conducting team training. To perform technical rescues safely and effectively coordinated efforts on the part of everyone are necessary. Personnel must know their individual role and their job within the team. Standard operating procedures or guidelines should clearly illustrate the roles and responsibilities for each position on the team, up to the Incident Commander's responsibilities. Most importantly, the team members must constantly retrain to further develop their teamwork skills to function as an efficient and effective unit.

# CHAPTER 8: TECHNICAL RESCUE TEAM EQUIPMENT

Technical rescue capabilities require standard equipment commonly carried on fire apparatus and specialized equipment that is not commonly carried. Most fire departments can provide basic rescue services with only a small investment in new equipment; however advanced rescue capabilities require a much larger equipment investment. This chapter discusses technical rescue equipment, personal protective clothing, and rescue vehicles.

## INITIAL EQUIPMENT

Emerging technical rescue teams should focus on mastering one area of technical rescue and acquiring the appropriate equipment necessary for this area of rescue before expanding to cover other areas. Many new teams focus on developing a rope rescue expertise first because the equipment required for rope rescue is more affordable than that required for other areas of rescue.

No matter which area of rescue you decide to address first, you should remember that some basic minimum equipment is necessary to safely execute a technical rescue. Appendix D lists the basic equipment necessary to start a rescue team. Prices are also provided to allow you to estimate how much of an investment you will need to make. Additional equipment that is suggested, but not required, for an emerging team is also listed.

Remember that some of the basic equipment you need to perform rescues may already be carried on your engines and trucks. Mechanical advantage systems can be rigged without carabiners or pulleys. Tripods for below grade rescues can be made by lashing two ladders together. Simple tools and simple physics principles, applied by trained rescuers, can overcome many obstacles to rescue. The Pyramids were built using these principles!

Teams performing advanced rescue skills can also utilize some very basic, inexpensive tools to safely accomplish a rescue. For example, trench rescues can be perform using wood shoring instead of special pneumatic or mechanical struts. Collapse debris can be lifted using jacks.

However, building wood shoring for a trench is laborious and time consuming, and carrying a large, heavy jack system capable of lifting over 100 tons may be impossible unless you have lots of available compartment space on your rescue vehicles. Special and more expensive tools such as pneumatic struts and air bags are preferable to use because they require less set up time than their counterparts mentioned above. Because some high-tech, specialized equipment is so expensive, you may consider purchasing it over a several year period so that costs can be spread out.

Ideally, equipment should be durable and have multiple uses. The U.S. Fire Administration's *Technical Rescue Technology Assessment* guide (refer to Appendix C) describes many types of technical rescue equipment you might consider purchasing when starting a new team.

## ALTERNATIVE SOURCES OF EQUIPMENT THROUGH A COMMUNITY RESOURCE PLAN

One of the options to alleviate the expense of setting up a technical rescue team is to develop a community resource plan (CRP), which is a plan to utilize equipment and personnel available within the community. A simple type of CRP is an arrangement with a lumber yard to provide wood for shoring in the event of a major collapse incident. There are innumerable types of equipment and personnel resources available in most communities.

Once you have identified the personnel and equipment resources that you will be able to provide from within your organization, you can then develop a CRP for the areas where you have insufficient resources. Next, look into your community to identify businesses, universities, suppliers, or individuals that may be able to provide you with the necessary resources to fill the "gaps." Sources of various types of resources are listed in Figure 8-1. It will be necessary to establish written agreements with these resource suppliers to ensure that they understand what will be expected of them in an emergency. In some cases, however, the suppliers may be unwilling to enter into an agreement with your department, but they are willing to help you if needed. Generally, even if

*Figure 8-1. Potential Resources Needed for*
*Technical Rescue Teams*

| Supplier | Resource |
|---|---|
| Construction/heavy equipment companies, state and local public works agencies | Backhoes, cranes, air compressors, dewatering pumps, dozers, loaders, welders, bobcats, generators, cherry pickers, tractor trailers, lighting, heavy tools, cutting and breaching equipment |
| Rental companies | Light tools, lighting, generators, air compressors |
| Lumber yards | Lumber, cutting equipment |
| Association of Engineers | Civil engineers, electrical engineers, fire protection engineers |
| Communications and warnings | Television stations, radio stations, ham radio groups |
| Emergency equipment suppliers | Sandbags, hazardous waste removal firms, vacuum trucks |
| Schools, churches, Red Cross, food suppliers | Disaster centers, food, and shelters |
| Funeral homes and medical examiners | Morgue services |
| Helicopters | Medevac, rescues, aerial photography, personnel and supply transport |
| Military/National Guard | Personnel, equipment |
| Transport companies | Equipment and supply transport, refrigerated trucks |
| Utility companies | Utility shut-off |
| Bottled water companies | Bottled water |

they are unwilling to enter into an agreement, if there is an actual emergency, they will be willing to help.

Some communities have published community resource directories that list resources within a community or even a large metropolitan area. This "database" of resources can be an important tool when an emergency strikes. The Greater Kansas City Metropolitan Region has established a booklet entitled the *Emergency Resources Catalog* under its "Plan Bulldozer Committee" which lists various resources and contacts within the region (refer to Appendix C for an address for the Plan Bulldozer Committee). Your list of resources should also include experts who may be of assistance such as civil engineers, communications specialists, fire protection engineers, etc.

## EQUIPMENT STANDARDS

When purchasing equipment, you should always ensure that the equipment meets basic standards. Many different organizations have established equipment standards including Underwriters Laboratory (UL), NFPA, American National Standards Institute (ANSI), American Society for Testing and Materials (ASTM), and National Institute of Occupational Safety and Health (NIOSH). These standards define performance measures, testing procedures, and safety and design features for equipment. Almost all rescue equipment must meet certain design standards, but the standards will vary depending on the type of equipment. Most equipment manufacturers should be able to tell you which standards their equipment meets.

## PERSONAL PROTECTIVE EQUIPMENT

Each discipline of technical rescue will require different personal protective equipment. The basic items usually include coveralls, helmet, firefighter's gloves or leather gloves, breathing apparatus, and protective clothing. Personal equipment you may consider purchasing for individual rescue disciplines is listed in Chapter 7.

You may want to obtain a copy of a free publication entitled *Protective Clothing and Equipment Needs of Emergency Responders for Urban Search and Rescue Missions* from the US. Fire Administration (refer to Appendix C for information on obtaining this manual).

## TECHNICAL RESCUE RESPONSE VEHICLES

A variety of types of technical rescue response vehicles are used across the country. Some departments purchase large squads dedicated as technical rescue units while others use a fire engine to carry rescue equipment. (Many areas of the country use different terminology to refer to rescue vehicles. Different parts of the country call these trucks rescues, squads, rescue squads, special service units, or heavy rescues. In this manual, the terms are used interchangeably to refer to large, dedicated technical rescue apparatus.)

No matter what type of vehicle you decide to use for technical rescue, remember that you will need a large amount of compartment space to carry the equipment needed for most technical rescue disciplines. Compartment spaces that are designed flexibly with adjustable shelves will allow you to modify compartments as necessary to fit new equipment. It is advisable to design new units that provide space to carry additional equipment as the team grows and technology changes. Many departments have made the mistake of purchasing rescue apparatus that is overloaded with equipment almost from the time it is delivered. Be sure you know what equipment you intend to carry on the vehicle and how much equipment the vehicle is rated to carry before you make purchases. A new NFPA standard (NFPA 1905) which will provide guidelines for special service vehicles, including rescue vehicles, is currently being developed.

Some of the technical rescue vehicle approaches you can consider are listed below. Pictures of various designs of rescue vehicles are included at the end of this chapter.

*Heavy Rescue Squads.* Many departments have heavy rescue squads designated for technical rescue. These are usually large fire-rescue apparatus with a body designed solely to carry rescue equipment. Some of these units contain a small water tank and fire pump; however, many do not and are used solely for rescue equipment.

There are many types of heavy rescue squads. Some squads are on commercial chassis or custom fire chassis. Crew members can ride in a cab area or in a walk through section, depending upon the design of the unit. Compartment bodies may be custom designed by the manufacturer, or have standard, beverage truck style compartment space.

Heavy rescue squads carry many types of equipment. Squads often have air supply systems for supplying air to pneumatic tools or refilling SCBA bottles. Many squads carry automobile extrication equipment. Technical rescue teams may store equipment on squads that perform other duties, such as fireground support operations, or they may have their own unit dedicated solely to their technical rescue mission. Some units will also attach special features to this type of vehicle such as a winch or crane, as pictured on one heavy duty squad at the end of this chapter. Heavy squads can cost between $125,000 and $500,000, depending upon the design and equipment. It is not unusual for a heavy rescue squad and all its rescue equipment to cost over a million dollars.

*Medium Duty Rescue Squads.* Medium duty rescue squads are generally units built upon medium duty truck chassis with utility bodies. Many technical rescue teams use medium duty squads to fulfill their needs. These squads carry less equipment than heavy squads but often serve the needs of technical rescue teams well. Medium duty rescue squads can cost between $50,000 and $150,000. Many teams have built their own medium duty rescue units with donated trucks from utility companies or public works departments.

A versatile type of medium duty squad unit used by several rescue teams across the country is a platform on demand (POD) unit. PODs are large storage containers that can be carried on a chassis, as pictured in this Chapter. A fire department can have several PODs holding different types of equipment (e.g. technical rescue, mass casualty, communications). Overall costs are reduced because the department needs only one chassis to transport any of these containers to the scene. Several FEMA US&R Task Forces use PODs because the containers are easily transported by aircraft or large trucks.

*Light Duty Rescue Squads.* Light duty rescue squads can range from four-by-four pickup trucks with utility compartments or caps, to ambulances and converted vans. A converted step van is a type of light duty squad pictured in this Chapter. Light duty squads are generally inexpensive units, though custom light duty trucks could cost up to $75,000. Generally, a light duty unit could be purchased for between $25,000 and $50,000. Many teams use donated light duty vehicles.

Light duty squads trade off the capacity to carry lots of equipment in return for reduced costs and a high degree of mobility, Many technical rescue teams use these vehicles because of their ability to travel off road or into areas that heavier apparatus cannot reach. Since many types of technical rescues take place in out-of-the-way areas, light duty units can be very versatile for technical rescue teams, especially those involved in rope rescue or water rescue.

Trailers. The least expensive method of transporting technical rescue equipment is a trailer. Trailers come in a variety of sizes and can carry large amounts of equipment. They are particularly good for carrying wood and shoring equipment for trench rescue teams. Some departments build shelves inside of trailers to create compartment spaces.

## RESCUE BOATS

A variety of types and designs of rescue boats are available to water rescue teams. Some are designed for use in calm water while others can be used in swiftwater, the ocean, or on ice. A comprehensive discussion of rescue boats and their capabilities is provided in the U.S. Fire Administration's *Technical Rescue Technology* Assessment manual (refer to Appendix C).

This New York City Fire Department vehicle is literally "maxed-out" with equipment for water rescue and other technical rescue emergencies.

This elaborate rescue vehicle is equipped with its own crane to remove concrete slabs and other objects at a collapse rescue incident.

Older engines and other vehicles can be refurbished for use by a technical rescue team.

A platform on demand (POD) system allows rescuers to stock multiple containers with equipment, but necessitates the purchase of only one transport vehicle. *(Courtesy Mike Tamillow)*

A simple step van can function as a low cost technical rescue equipment transport vehicle.

A trailer is a low cost means to transport large amounts of equipment. Sometimes, a rescue agency can arrange for transport units to be donated, as this one was.

Technical rescue vehicles should be designed with large equipment compartments.

# CHAPTER 9: TECHNICAL RESCUE INCIDENT COMMAND AND SAFETY

**M**ost fire and rescue agencies are familiar with the Incident Command System (ICS) and use it regularly to manage operations and resources at emergency incidents. Utilization of ICS at a technical rescue incident or training event, just like any major fire or EMS incident, is essential. ICS is an organizational tool that provides a standard system to manage resources, coordinate operations among units, and monitor scene safety. Technical rescue fits within the standard ICS approach to incident management. This chapter discusses the application of ICS to technical rescue incident command, the specific roles assigned to command personnel, and the utilization of a technical rescue safety officer,

## TECHNICAL RESCUE INCIDENT COMMAND SYSTEM POSITIONS

It is beyond the scope of this manual to describe the philosophy and structure of the ICS in detail. This chapter assumes that the reader has a basic understanding of how the ICS works. (Entire courses are available on the fundamentals of ICS. Readers for whom this is new material should consult ICS training materials and additional resources, including the text *Fire Command* listed in Appendix C).

A fundamental tenet of the ICS is that it is flexible in its application and it can be used for all types and sizes of incidents. This is the controlling principle when discussing how ICS applies to

*Figure 9-1. Incorporation of a Technical Rescue Branch Into the ICS Structure*

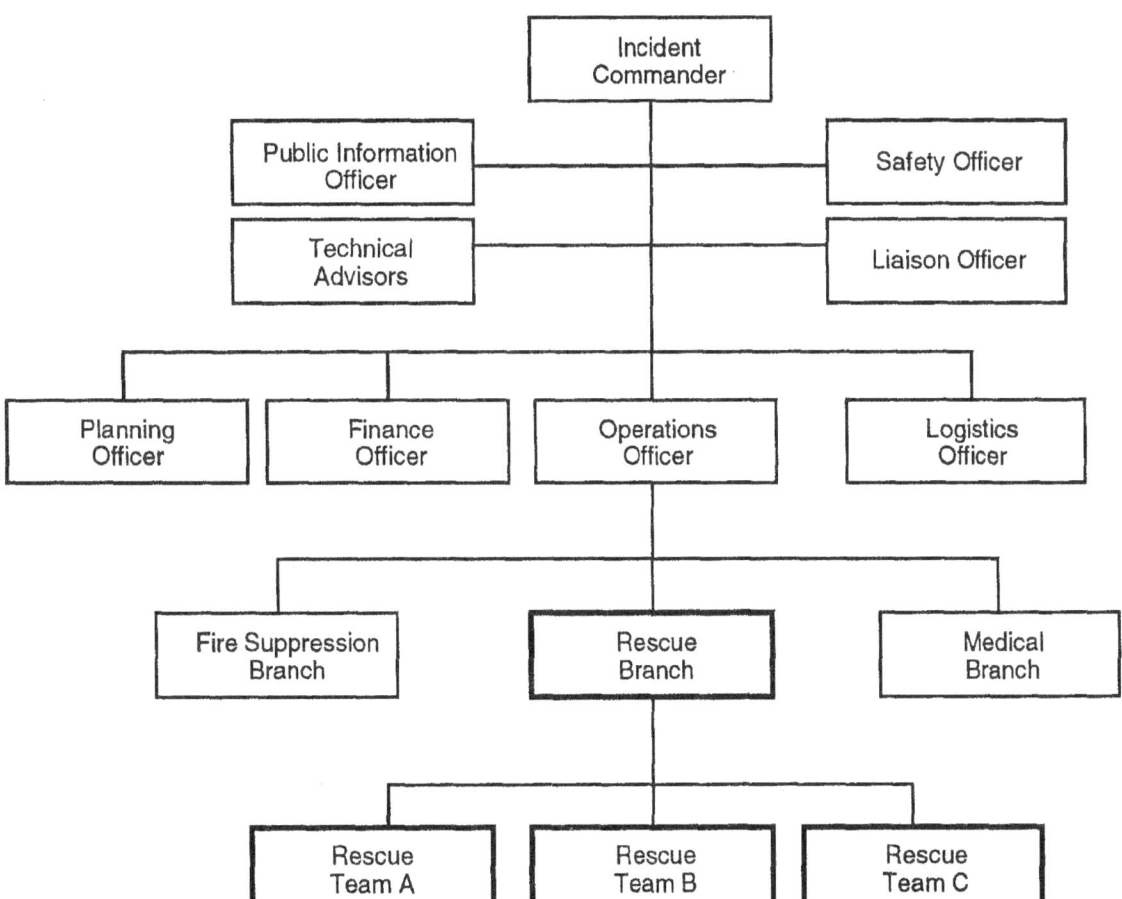

technical rescue incidents. In essence, the fundamental ICS structure remains the same for all types of incidents - including fire, EMS, technical rescue, or hazardous materials incidents - allowing for an Incident Commander with four sections below (operations, logistics, planning, and fiance as shown in Figure 9-1). However, depending on the size and nature of the situation, only the components that are needed will be established for each incident.

At large scale incidents, the Technical Rescue can be assigned as a branch within the Operations Section (see Figure 9-2). This part of the organization would be supervised by a senior member of the technical rescue team. In a "branch" organization, specific responsibilities within the technical rescue team could be established as "group" functions. In smaller scale incidents, it could be assigned as a "group" or "sector."

Generally, you should consider expanding the ICS to include a technical rescue branch or group when any of the following conditions apply:

• The nature of the technical rescue(s) involved is too complex or hazardous to be carried out safely and effectively by non-specialist personnel, thereby requiring the assistance of a special rescue team.

• The rescue is projected to take more than an hour to effect.

• There are multiple types of fireground operations occurring simultaneously (i.e., hazmat, EMS, suppression) in addition to the technical rescue(s).

• The span of control of an Operations Officer would be exceeded by attempting to implement a technical rescue operation.

*Figure 9-2. Technical Rescue Branch Command Structure*

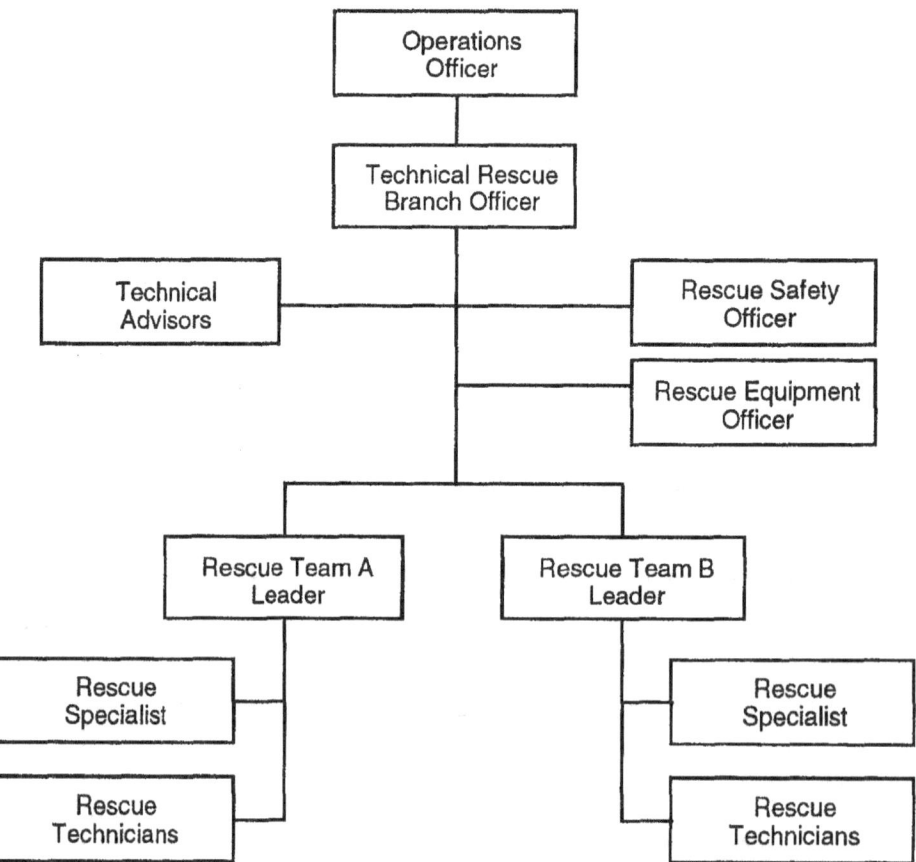

The Operations Section Chief will have the responsibility to coordinate activities where simultaneous rescues must be integrated with functions such as hazardous materials control, fire suppression, jurisdiction receives the assistance of a mutual aid technical rescue team. In these situations, the Incident Commander should assign the technical rescue team to work together and function as a

*Figure 9-3. Technical Rescue Command Position Roles*

| ICS Position | Role/Responsibilities |
|---|---|
| **Technical Rescue Branch Officer**<br>*Reports to: Operations Officer* | 1. Coordinates the overall rescue operations ongoing at any time<br>2. Responsible for implementation of specific technical rescue objectives as defined by the Operations Officer |
| **Rescue Team leader**<br>*Reports to: Technical Rescue Branch Officer* | 1. Supervises rescue efforts of a specific nature or at a specific geographic location<br>2. Responsible for smooth execution of particular rescue functions |
| **Rescue Safety Officer**<br>*Reports to: Technical Rescue Branch Officer* | 1. Monitors rescue team activities for safe operating practices<br>2. Is authorized and *required to* halt any activities which pose a danger to rescuers, victims, or bystanders<br>3. Coordinates actions with Incident Safety Officer |
| **Rescue Equipment Officer**<br>*Reports to: Technical Rescue Brunch Officer* | 1. Stages and issues technical rescue equipment<br>2. Coordinates with Logistics Section to obtain necessary additional specialized equipment<br>3. Maintains accountability for rescue equipment |

and EMS. The responsibilties of individual officers below him or her are described in Figure 9-3.

In many rescue situations, the Incident Commander may not have the level of training or specific expertise to directly supervise technical rescue operations. By assigning a qualified supervisor to manage the Technical Rescue Branch, the Incident Commander retains the responsibility for managing the overall incident while placing a capable individual in the position to ensure that the technical rescue functions are conducted safely and effectively. This situation may arise when one

unit within the organization. The Incident Commander is still in charge of the incident, but a qualified individual is in charge of technical rescue.

All emergency response personnel should be familiar with the Incident Command System design and have a thorough understanding of their individual roles within the system.

## TECHNICAL RESCUE SAFETY

Technical rescue incidents, even more than many other types of incidents, can pose significant hazard to rescuers as well as the individuals they are attempting to rescue. Accordingly, it is important for the Incident Commander to assign an Incident Safety Officer to monitor scene safety,, verify that safe and proper procedures are used at all times, and notify the Incident Commander of any safety concerns. The technical rescue team should have one or more individuals within the team who are qualified to act as safety officers. In certain situations, the Incident Safety Officer would retain the overall responsibility for incident safety while the technical rescue team's own safety officer would oversee the technical rescue operations. In this case the Technical Rescue Safety Officer(s) (who should be trained to the technical rescue level) should be assigned to work under the Technical Rescue Branch Officer and coordinate activities with the Incident Safety Officer.

Figure 9-4 outlines some of the hazards which responders may have to face on different types of incidents. This is not meant to be an all-inclusive list but merely an example of factors that an

*Figure 9-4. Technical Rescue Hazards*

| Incident Type | Some Potential Hazards |
|---|---|
| **Confined Space** | • Hypo-/hyperthermia<br>• Oxygen deficient and toxic atmospheres<br>• Exposed utilities<br>• Entrapment |
| **Water** | • Hypothermia<br>• Electrocution<br>• Floating hazards (e.g., logs)<br>• Low-head dams (drowning machines)<br>• Entrapment in submerged hazards<br>• Drowning<br>• Swift currents<br>• Boats |
| **High-Angle** | • Hypo-/hyperthermia<br>• Falling debris<br>• Rope system or equipment failure<br>• Falls<br>• Improper rigging |
| **Trench/Cave-in** | • Exposed utilities<br>• Secondary collapse<br>• Hypothermia<br>• Oxygen deficient atmospheres |
| **Industrial** | • High noise environments<br>• Heavy machinery<br>• Exposed utilities<br>• Hazardous material releases<br>• Falls<br>• Confined spaces<br>• Stored energy release |
| **Building Collapse** | • Entrapment by falling debris/shoring failure<br>• Secondary collapse<br>• Hypo-/hyperthermia<br>• Oxygen deficient atmospheres<br>• Exposed utilities<br>• High noise environments |

Incident/Sector Safety Officer must consider when monitoring a technical rescue evolution.

## OUTSIDE SPECIALISTS

At certain technical rescue incidents such as building collapses, the Incident Commander may need to request the assistance of an outside technical specialist to review the scene, provide additional safety observations, and evaluate the Incident Commander's action plan. Technical specialists, such as structural engineers, may not be available from within the emergency service community. Another example might be the use of plant personnel to describe hazardous environments or to map out confined spaces.

When calling upon outside experts, the Incident Commander must have confidence that the individual is competent and qualified to make assessments of the situation and provide good advice. It is desirable to have established relationships with individuals who are known to have the desired expertise and to understand the skills and capabilities of the technical rescue team.

## TRAINING SAFETY

The Incident Command System and safety officers in particular should be used for all rescue training evolutions. Training activities often involve the same types of hazards as real incidents; however, training exercises provide the opportunity to identify, evaluate, and take appropriate steps to control the hazards before initiating any activities. Unfortunately, there are numerous examples of training exercises that resulted in injuries or fatalities because adequate safety and incident command measures were not taken. Technical rescue training is no place for a cavalier attitude. Failure to conform to proper training safety protocols is inexcusable and may result in personal or departmental liability judgments.

All technical rescue training evolutions should be performed under the rigorous scrutiny of a safety officer operating within an ICS. Since all technical rescue operations will take place in real life using an ICS, it is logical to use an ICS for the training evolutions. Aside from maintaining the high safety standards, the increased exposure to the application of the ICS is valuable training in itself.

Obviously, training evolutions, like small scale incidents, will not require the fully expanded ICS that a real incident might generate; however, at a minimum there should always be an Incident Commander and an Incident Safety Officer.

# CHAPTER 10: TECHNICAL RESCUE MUTUAL AID

**M**utual aid is a very practical and cost effective method to manage fire, EMS, or even technical rescue service response. Mutual aid is cooperative effort by participating jurisdictions to help one another in day to day or major responses. The dynamics of mutual aid agreements have changed dramatically over the past few years. Large scale disasters in highly populated urbanized areas of the United States, including the Oklahoma City bombing, have renewed interest in the coordination of multi-agency forces in the handling of emergency incidents. These events have made emergency managers realize that the management of mutual aid forces are essential to handle a large-scale disaster.

Multi-agency mutual aid technical rescue teams have been started in many communities across the country. Because of the low frequency of technical rescue incidents compared to fire and EMS incidents, many communities have found that it is difficult to justify a rescue team in their own jurisdiction but important to have the capability when it is needed. In many cases, communities have set up regional teams that are available to respond to any participating jurisdiction through a mutual aid system.

This chapter discusses some of the considerations that are necessary when forming an inter-agency or regional mutual aid technical rescue response system.

## CONSIDERATIONS WHEN FORMING A MUTUAL AID TEAM

Below is a list of some considerations you may want to make before starting a technical rescue mutual aid team. Note that technical rescue mutual aid systems are very similar to firefighting mutual aid systems many communities already have established.

- There must be a recognition of a common need among a group of jurisdictions for mutual aid services. The first factor in determining if a shared resource system is appropriate is to determine if both (or all) of the communities will mutually benefit from the sharing of resources.

- The agencies and communities must be willing to participate in a joint venture. It is essential to have or to establish a high level of compatibility and trust among the agencies that are attempting to join forces. Unfortunately, the priorities, cultures, and traditions of many organizations can provide insurmountable obstacles in establishing effective systems. These obstacles can be at any level, from elected officials, to managers and chief officers, to labor organizations or rival companies.

- There must be agreement on the funding approach or formula for the mutual aid system. Different agencies may have different abilities to fund shared costs. There are numerous ways to work out creative reimbursement systems that attempt to reach equity in sharing these costs, including consideration of in-kind services such as vehicle maintenance, communications and dispatching, training, and other functions. As long as the parties agree to what is fair and reasonable, this problem can be resolved in a host of ways. The key is to establish an understanding or agreement that is acceptable and approved by the participating jurisdictions.

- The participating agencies must agree to support a system of standard operating procedures that will apply in any situation where a multi-agency team would respond and operate.

- Geography can also impact on the feasibility of a multi-agency program. Response time and distance must be considered for functions that require rapid intervention.

- One agency in a region could already have a technical rescue capability which could be available to other jurisdictions under a contract or multi-agency agreement. The agency that provides this service may seek monetary support form the other jurisdictions, or it may choose to provide this service at no charge.

## MUTUAL AID SYSTEM DESIGNS

There are three common designs of regional mutual aid structures: the "pooled resources design," the "shared specialty design," and the "standalone design." The pooled resources design refers to a system where several departments or jurisdictions provide personnel, resources, and/or funding to organize and operate a team. The shared specialty design refers to a system where individual jurisdictions provide individual specialty teams, such as one providing a dive rescue team while another provides a confined space rescue team. The standalone design refers to a system where one jurisdiction has a specialized capability and may charge other jurisdictions for its services as a means of recovering the cost.

Any of these designs can incorporate itself into a tiered response, as described in Chapter 1, in which all personnel in a region have basic (awareness level) training in technical rescue and know when to call for the regional response team that has advanced training and capabilities for complex situations.

## COMPONENTS OF A REGIONAL TIERED RESPONSE SYSTEM

An effective regional or mutual aid team utilizing a tiered response must be able to integrate its operations within the incident management organization established by the first responders. Several organizational components are important to the success of this type of system.

- There must be a commonly accepted *standard operating procedure* that guides a tiered response. Initial responders must be trained in the detection and assessment of the situation so that proper notification and activation of the special team is carried out.
- *All workers must be familiar with SOPs, tools, equipment, and safety measures.* The key component of an effective regional team concept is the ability of the specialized team to integrate its operations within the scene organization established by first responders.
- .A standard approach to *incident command* provides the framework for different agencies to integrate their operations at an incident. Its proper application and the discipline with which emergency workers utilize this process will make or break the effectiveness of the regional team concept.
- There must be an effective radio *communications system* which allows all responders to access the same channel (or a patched channel) at an incident.
- *Interagency training* several times a year will prepare rescuers for an interagency response. In addition to rescue team training, all responders

should be trained as to what to expect when the regional rescue team arrives at an emergency scene. Preparedness is vital to making a regional system work efficiently.

## INCIDENT COMMAND AND INTERAGENCY RESPONSE

The agency having jurisdiction always maintains the overriding authority and responsibility for control of the incident. The agency having jurisdiction may delegate the management of operations to a better equipped or more qualified responding agency, but it is ultimately responsible for what happens in its own jurisdiction. The incident command structure should be outlined in a mutual aid agreement and it should specify how the system will be structured, how each agency will interact with another within the system, and who is ultimately in charge.

The overriding intent of the incident command system is to provide an integrated management system to handle large-scale events. This structure should be standardized among all the jurisdictions participating in mutual aid. It may be designed around a unified command structure which is used when different agencies are responsible for different aspects of the situation. In most technical rescue mutual aid situations, a technical rescue team would normally be assigned as a group within the incident management system. The Technical Rescue Team Leader would assign different functions to individual team members. (Refer to Chapter 9 for a more detailed discussion of technical rescue incident command.)

## MUTUAL AID AGREEMENTS

Most governmental agencies are authorized to establish mutual aid relationships or cooperative agreements with other agencies including local, state, Federal, military, private, or international entities. The specific features of an individual agreement are defined by the needs and desires of the participants. The following is a list of issues that each agreement should address:

- Agencies involved in the agreement
- Jurisdictional authority and service areas
- Descriptions of services provided
- Activation or requesting process
- Reporting and documentation process

- Reimbursable costs or fees for service
- Refusal options
- Liability provisions
  - Employee injuries &workers compensation
  - Damaged equipment
  - Damaged personal property
  - Legal costs
- Term of the agreement
- Termination provisions

Mutual aid agreements provide the legal authorization for a jurisdiction's resources to be utilized outside of its boundaries. They should be supported by SOPs and guidelines that define how operations will be conducted. An operational guide of this type should address the following provisions:

- Standard description of vehicles and equipment
- Staffing levels and training competencies
- Response protocols (unit assignments)
- Command structure (including passing and assuming command)
- Unit identification and personnel designations
- Radio communication procedures and nomenclature
- Equipment placement and report-in procedures
- Field operating procedures
- Personnel accountability, rehabilitation and medical monitoring
- Incident reporting and press releases
- Joint training evolutions

Refer to Appendix A for sample mutual aid agreements.

## ROLES AND RESPONSIBILITIES OF THE FIRST RESPONDER

In most technical rescue incidents, a technical rescue team unit will not be the first unit to arrive on the scene. Instead, the technical rescue team would be dispatched to support fire suppression and/or EMS units that would be the first responders to the technical rescue emergency incident.

The first responder units may only have the training to perform a basic assessment of the situation and to initiate basic emergency actions, but often their role is crucial in beginning to stabilize the incident and preparing for the arrival of the technical rescue team. In this case, they must know what to do and what not to do until the rescue team arrives. The degree of interaction that takes place between first response units and incoming technical rescue teams will have a direct impact upon the overall effectiveness of the management of the scene.

First responders may have a tendency to minimize pre-arrival radio communications with an incoming technical rescue team for fear that they will appear less knowledgeable than the technical experts regarding the particular situation. (This may be more common when the first responders are not familiar with the incoming rescue team.) It is important for first responders to be trained in sizing-up the situation and reporting their observations to technical rescue teams while they are en route or when they arrive on the scene. First responders should be able to report the nature of the incident and the level of response that is necessary, as well as the potential for secondary collapse and other identified hazards.

In many cases, the primary responsibilities of the first responders are to recognize the characteristics of the call, identify the type and level of specialist response that is necessary, cordon off the hazard zone, and ensure the safety of the first responders by minimizing their risk of exposure. There are situations, however, where the first responders may take important actions to rescue or protect endangered persons within the limits of their capabilities.

The time factor may ultimately dictate the degree of involvement that first responders will have in mitigating a situation. The rescue of victims is the first priority of any emergency responder. It is critical for first responders to understand the hazards presented by the incident before committing themselves to a rescue attempt.

Training evolutions are extremely helpful in training first responders. Simulations can be designed in which first responders are required to observe and report as much of the relevant information as possible. These simulations can utilize pictures or video to present the information. First responding officers are given the task of reporting what they see and also determining what they can or should do in each situation. The first responders can also simulate a verbal report of their observations and the technical rescue team could practice making a "size-up" without seeing the visual. Through this type of training, the workers are able to develop effective working relationships. If members of different agencies are expected to work as an effective unit, they must be able to develop a degree of familiarity and confidence in each other. Dialogue is a key to initiating a successful operation and ensuring that a proper execution of the mission will follow.

## CONSIDERATIONS WHEN RECEIVING TECHNICAL RESCUE MUTUAL AID ASSISTANCE

The first responders to a technical rescue emergency will determine if they are capable and equipped to handle the incident or if they will need the assistance of a specialty rescue team. The advanced team may be responding from another jurisdiction and there may be a substantial delay before they arrive.

Below is a list of actions that first responders should consider while waiting for a technical rescue team to arrive:

- *Collect as much information* about what happened as possible. Find out how the accident occurred, and how many victims need assistance. If victims are missing, find out where and when they were last seen. This information should be relayed to the incoming rescue team or be ready when they arrive.

- *Establish communications with incoming mutual aid units,* This will allow the on-scene units to convey vital information to incoming units and it will allow the incoming units to give instructions to the on-scene units for

actions they should take until the incoming units arrive.

- *Secure the scene and remove hazards.* Cordon off the scene and remove any bystanders who are in the way of rescuers. To the extent possible, remove any hazards or at least make note of any hazards so that incoming rescuers can be made aware of them.

- *Prepare an arrival area* for incoming mutual aid units. Find out how much space they will need and **the** number of vehicles they are bringing. If necessary, designate a staging area.

- *Set up any support resources.* It may be possible to set up support resources that will be necessary to handle the incident before the rescue team arrives. Call for additional personnel to support the rescue team, set up lighting, request an air support unit, have food, drinks and shelter space ready for rescuers, obtain additional wood for cribbing, call for heavy equipment and equipment operators, etc. The incoming units can advise while en route of special resources they anticipate needing upon their arrival.

- *Have EMS personnel standing by.* In extended operations, it may be advantageous to have an emergency physician on scene to assist. Notify the hospital of the situation, number of victims involved, and the nature and extent of injuries. Request a medevac helicopter if needed.

- *Make other notifications as necessary.* Other agencies may have to be notified of this incident including your state OSHA office.

- *Establish the command system. Any* advanced rescue operation should be conducted within the framework of the incident command system (refer to Chapter 9). The command system can be in place before outside rescuers arrive.

# CHAPTER 11: THE NATIONAL URBAN SEARCH AND RESCUE RESPONSE SYSTEM

**I**n the past few years, there has been widespread publicity about the Federal Emergency Management Agency's (FEMA) Urban Search and Rescue (US&R) program. This program has been used to provide expert technical rescue teams after major disasters including Hurricane Andrew, Hurricane Iniki, the Northridge, California Earthquake and the Oklahoma City Federal building bombing. Eleven US&R teams, called Task Forces, were dispatched to Oklahoma City in April 1995 to assist with search and rescue at the collapsed Federal building.

This chapter provides an overview of the FEMA Urban Search and Rescue program and shows how this Federal program interfaces with local fire departments in times of disaster. It also discusses the process by which you could request a FEMA rescue team in a real disaster.

## THE FEMA US&R SYSTEM

The Federal Emergency Management Agency coordinates the National Urban Search and Rescue Response System. This system currently comprises a network of 25 specially equipped and trained US&R Task Forces from localities across the country which can be requested to provide assistance at large-scale disaster (see Figure 11-l). The Task Forces in the system are deployed by FEMA when requested by a state or locality. Other components of the system include Incident Support Teams (ISTs) and technical specialists in the urban search and rescue field.

*Figure 11-l. Locations of FEMA US&R Tusk Forces*

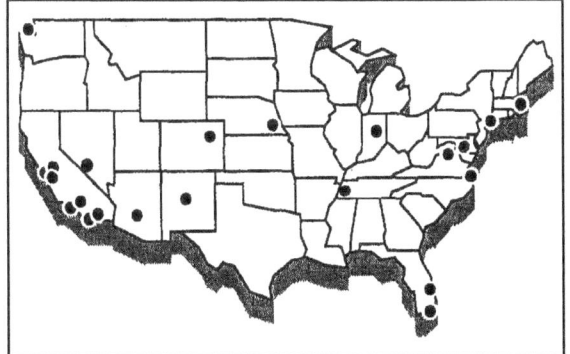

The FEMA US&R Response System is designed to provide a coordinated response to disasters in the urban environment. Special emphasis is placed on the capability to locate and extricate victims trapped in collapsed structures, primarily of reinforced concrete construction. The following list illustrates some the principal Task Force operational capabilities:

- *Physical and technical search and rescue operations in damaged collapsed structures*
- *Emergency medical care to Task Force response personnel*
- *Emergency medical care to the entrapped victims*
- *Reconnaissance to assess damage and needs and provide feedback to local, state, and Federal officials*
- *Assessment/shut off of utilities to houses, buildings*
- *Hazardous materials surveys/evaluations*
- *Structural/hazard evaluations of government municipal buildings needed for immediate occupancy to support disaster relief operations*
- *Stabilizing damaged structures, including shoring and cribbing necessary to operate within a structure*

Each FEMA US&R Task Force contains 62 specially trained personnel. A Task Force is designed to be logistically self-sufficient for the first 72 hours of operation and to be able to function for up to 10 days before being replaced by another Task Force. The 62 person Task Force divides into two groups, each of which operates in 12 hour shifts on the scene. One group works while the other rests. All Task Force members must be sufficiently cross-trained in their search and rescue skill areas to ensure depth of capability and integrated Task Force operations. By design, there are two personnel assigned to each identified Task Force position for the rotation and relief of personnel. This allows for round-the-clock Task Force operation.

Each Task Force brings its own equipment cache to the scene. This cache is described in detail later in this chapter. The Task Force functional organization, operating procedures, and associated terminology are compatible with the National Interagency Incident Management System.

Each Task Force is designed to be able to be rapidly deployed in an emergency. All members must meet a six-hour window for mobilization. Depending on the location of the disaster, a Task Force will respond to the scene either by ground using its own trucks, or via a military or civilian aircraft. In general, an initial Task Force can be at the scene of a disaster within 24 hours anywhere in the contiguous states.

When the Task Forces are not on a FEMA national response, they function as technical rescue teams in their own communities and, in many cases, serve as regional or state-wide rescue teams.

## Task Force Component Teams

Disaster response experience has demonstrated the need for expertise and capabilities in four areas: search, rescue, medical, and technical assistance. To address this need, each of FEMA's US&R Task Forces is divided into four teams with skills in each of these areas.

**Search Team.** The primary focus of the Search Team. is to locate live victims trapped in collapsed structures. The team must provide canine, electronic and physical search strategies, and other search tactics and techniques to locate trapped victims.

**Rescue Team.** The primary responsibilities of the Rescue Team are the evaluation of compromised areas, structural stabilization, breaching and site exploration, and live victim extrication.

**Medical Team.** The Medical Team is designed to provide sophisticated and possibly prolonged pre-hospital and emergency medical care at rescue sites. The medical personnel are also responsible for minimizing health and safety risks, critical incident stress debriefing, caring for sick team members, and providing treatment for Task Force personnel exposed to hazardous materials. In addition, the medical personnel must be capable of providing treatment to the Search Team canine.

The treatment priorities for the Task Force Medical Team are:

*First:* Task Force members and support personnel

*Second:* Victims directly encountered by the Task Force

*Third:* Task Force search canine

*Fourth:* Other victims as possible (it is not the intent of this team to be a free-standing medical resource at the disaster site)

Local medical systems are the primary providers of general medical care to disaster victims. It is recognized that the Task Force Medical Team may have to "hand off" a potentially unstable patient to a less equipped interim level of medical provider for transport to definitive care. This is considered to be standard practice under the circumstances of disaster operations,

Technical Team. The Technical Team is comprised of specialists that support the overall search and rescue mission of the Task Force. The primary responsibilities of the Technical Team are to provide an evaluation of hazardous or compromised areas, structural assessment, stabilization advice, and hazardous materials monitoring. The team also functions as the liaison with local emergency responders, coordinates communications and logistics, and documents the incident for the Task Force.

## US&R Task Force Equipment Cache

Each Task Force is supported by a comprehensive equipment cache that weighs over 30,000 pounds and which allows the Task Force to be self-sufficient for immediate operations.

A US&R Task Force equipment cache includes over 1,200 items. The following illustrates the comprehensive nature of a US&R equipment cache:

*Rescue Equipment Group*
* Electric generators
* Pneumatic air compressors
* Power tools
* Hand tools
* Electrical equipment
* Lighting
* Maintenance items
* Rope/rigging equipment
* Safety equipment
* Search cameras
* Seismic/acoustic listening devices

*Medical Equipment Group*
* Medicines
* IV fluids/volume expanders
* Immunizations/immune globulin
* Canine treatment
* Airway equipment
* Eye care supplies

- IV access/administration
- Patient assessment items
- Patient immobilization
- Patient/personal protection
- Skeletal care items
- Wound care items
- Cardiac monitoring equipment

*Communications Equipment Group*
- Portable radios
- Charging units
- Telecommunications items
- Repeaters
- Accessories
- Batteries
- Power sources
- Small tools

*Technical Equipment Group*
- Structures Specialist items
- Tech Info Specialist items
- Hazmat Specialist items
- Tech Search Specialist items

*Logistics Equipment Group*
- Water/fluids
- Food
- Shelter items
- Sanitation equipment
- Personal safety gear
- Administrative support items
- Personal gear

## THE INTERFACE BETWEEN A LOCALITY AND A FEMA US&R TASK FORCE

The sudden devastation in Oklahoma City following the bombing of the Federal building created a disaster scene which had the potential to overwhelm local response resources. Within hours of the incident, there was an outpouring of support from fire departments, EMS departments, hospitals, and other groups around the state and region.

After the patients who could be immediately removed from the structure were transported to hospitals, the fire department began to address the massive tasks of stabilizing the building, removing debris, and searching for other victims. FEMA's US&R Program provided supplemental resources to help the locality deal with these tasks. This section discusses how a state or locality interfaces with FEMA's US&R program and other Federal disaster programs. It also describes the process for requesting the assistance of a FEMA US&R Task Force.

## The Federal Response Plan

The Federal Response Plan is the Federal government's plan of action for responding to major disasters. It addresses the consequences of any incident or emergency situation in which there is a need for Federal response assistance. FEMA's US&R program is one component of the Federal Response Plan.

The Plan is applicable to natural disasters such as earthquakes, hurricanes, typhoons, tornadoes, and volcanic eruptions; technological emergencies involving radiological or hazardous materials releases; and other incidents requiring Federal assistance which fulfill the following criteria:

- *The State and local response capabilities are overwhelmed;*
- *The State government requests Federal assistance; and*
- *The President formally declares that a disaster or emergency has* occurred, *activating the disaster assistance authority outlined in The Robert T. Stafford Disaster Relief and Emergency Act, Public Law 93-288, as amended.*

The Plan describes the basic mechanisms and structures by which the Federal government will mobilize resources and conduct activities to augment State and local response efforts. Federal assistance is provided to the affected State under the overall coordination of a Federal Coordinating Officer (FCO) appointed by the Director of the Federal Emergency Management Agency on behalf of the President. The Plan groups the types of Federal assistance which a State is most likely to need under 12 emergency support functions (ESFs) (see Figure 11-2). A Federal agency is assigned to coordinate each ESF. Each agency has been selected based on its legal authorities, resources, and capabilities. Other agencies have been designated as support agencies for one or more ESFs. In the Federal Response Plan, urban search and rescue is ESF-9.

## The Process for a State or Locality to Request Assistance from a FEMA US&R Team

States or localities can request Federal emergency assistance during a time of disaster. The process for requesting assistance from a Federally-sponsored US&R Task Force is similar to any Federal emergency assistance request originating

*Figure 11-2. Emergency Support Functions*

| FUNCTIONAL ANNEXES | | |
|---|---|---|
| Emergency support Function | Annex | Primary Agency |
| ESF-1 | Transportation | Department of Transportation |
| ESF-2 | Communications | National Communications system |
| ESF3 | Public Works and Engineering | Department of Defense (Corps of Engineers) |
| ESF-4 | Firefighting | Department of Agriculture (Forest |
| ESF-5 | Information and Planning | FEMA |
| ESF-6 | Mass Care | American Red Cross |
| ESF-7 | Resource Support | General Services Administration |
| ESF-8 | Health and Medical Services | D.H.H.S., U.S. Public Health Service |
| ESF-9 | Urban Search and Rescue | FEMA |
| ESF-10 | Hazardous Materials | Environmental Protection Agency |
| ESF-11 | Food | Department of Agriculture |
| ESF-12 | Energy | Department of Energy |

at the state or local level. State emergency coordinators are usually familiar with this process, but below is an overview of the process for requesting the assistance of a FEMA US&R Task Force.

A formal request for assistance must be forwarded to FEMA by the Governor of the state, who may be acting on the formal request of local authorities. The Federal government must authorize assistance before Federal support is given. The authorization process can take a few hours to several days, depending on the nature of the emergency. The request must be made formally in writing, but an initial verbal request followed by a written request later can be accepted when there is a sudden onset emergency. Requests for Federal assistance should specify the type of resources desired (e.g. US&R assets). If the request for US&R Task Force assistance is approved, FEMA will activate its US&R emergency support function, ESF-9. FEMA then dispatches an Incident Support Team

(described further in the next section) and selected US&R teams to the scene as deemed necessary.

## US&R Incident Support Team (IST)

An IST consists of highly qualified rescue experts who are readily available for rapid assembly and deployment to a disaster area. Its primary responsibility is field coordination of US&R assets dispatched in support of a local jurisdiction. Upon request of US&R assistance, the IST is immediately dispatched to a disaster scene and in some cases may be the first arriving US&R support. It does not directly manage State or local disaster response activities, but assists local authorities. The IST coordinates all Task Force activities prior to their assignment to state/local jurisdictions. Other responsibilities of the IST include:

- Executing the mission assigned by FEMA and the state or locality

- Coordinating logistical support for the Task Forces, including resupply and coordinating transportation through other Federal departments/agencies

- Ordering additional resources

- Coordinating actions with other ESFs in the field (e.g. ESF-8 for Disaster Medical Assistance Teams and Disaster Mortuary Teams)

- Providing input for the situation report, incident action plan, briefings, resource status reports, historical documentation, and other information management

- Developing demobilization plans and executing demobilization of Task Forces

- Collecting information from the US&R liaisons, Task Force support specialists and US&R technical specialists

- Ensuring that personnel processing procedures are followed, and that personnel well being is assessed

- Providing cost estimates on US&R operations

## 1ST Organizational Structure

The initial response of a US&R Incident Support Team will be comprised of the following nine core positions as shown in Figure 11-3.

Initially the Incident Support Team is mobilized with the nine personnel described above. As the

*Figure 11-3. Initial Incident Support Team (IST) Response Unit Organization*

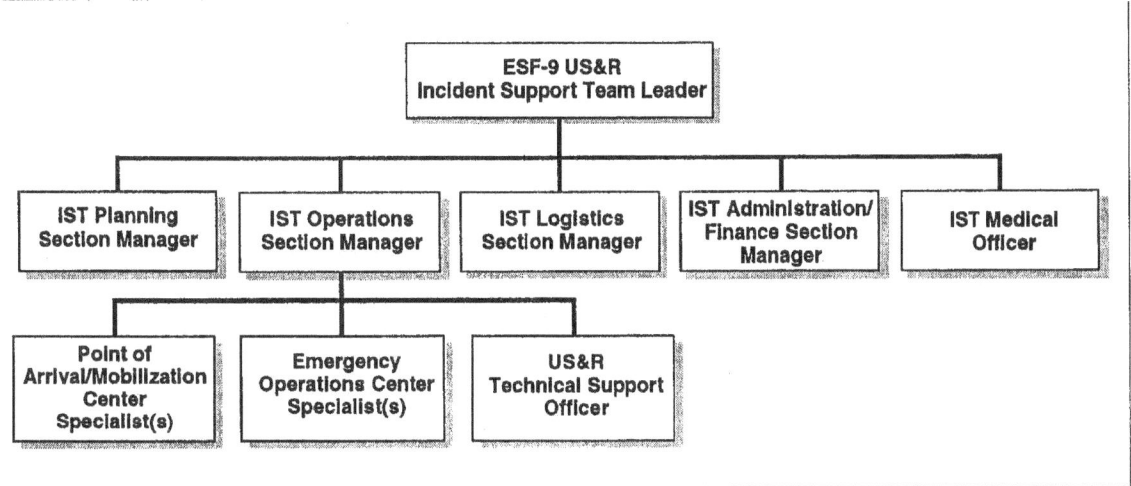

complexity and duration of an event escalate, it may be necessary to augment sections of it as the duties and responsibilities of the team expand. Figure 11-4 illustrates the organizational structure of a fully staffed Incident Support Team.

## IST Operations

Upon arrival of the IST, the IST Leader will make assignments and prepare for IST personnel to deploy to the field. Prior to deployment, IST personnel receive a briefing, including initial situation assessment, information about their Federal and State points of contact at the various facilities, mission orders, and relevant procedures.

The IST will dispatch personnel to the Task Force Point of Arrival, Mobilization Center, local Incident Command Post(s) or other areas in order to provide support and coordination with facility managers. IST personnel are familiar with the supporting Federal agencies. For instance, they will coordinate with the Department of Defense

*Figure 11-4. Expanded IST Response Unit Organization*

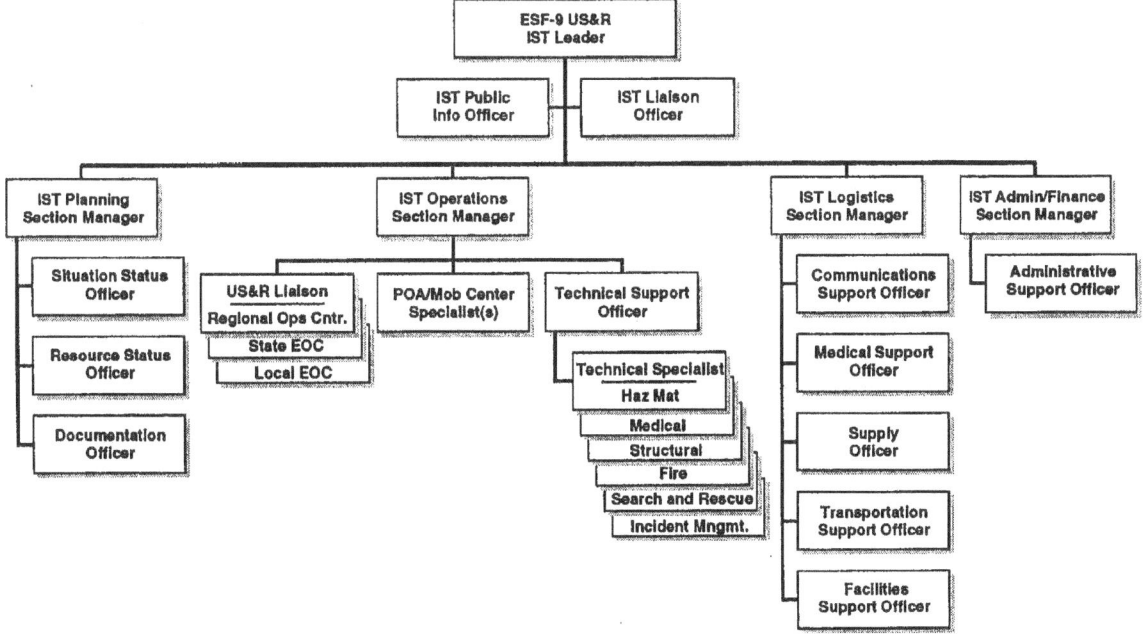

representatives at the Point of Arrival to arrange Task Force ground transportation, logistical support for Task Force personnel needs, briefings, support for off-loading equipment, and other activities to expedite the movement of the Task Forces to the next assigned facility.

The IST specialist assigned to the Federal Mobilization Center (set up near the incident) will work with the assigned Federal agency manager to ensure that billeting and food service is in place for the Task Forces. The IST specialist will work with ESF-1 to arrange transportation to and from the incident area for the Task Forces. The specialist will collect current data and will brief the Task Force Leaders and management staff. Other IST personnel may be dispatched to Federal, State, and local emergency operations centers in order to provide technical information on the US&R Task Force capabilities and to coordinate the appropriate application of these resources. IST personnel assist in planning US&R operations and develop requests for additional resources and forward them to FEMA Headquarters.

## Command and Control of a US&R Task Force

Management of the US&R Task Forces changes during the initial operations. The development of the national Urban Search and Rescue Response Program is founded upon the Task Force being a response capability for the requesting jurisdiction. During the notification and initial deployment phase, FEMA Headquarters maintains overall command and control of the teams and their assets. The Task Force is under control of FEMA, through the Incident Support Team, while en route to the affected locality. Once on scene and assigned to an incident/location, the Task Force reports to the authority of the local jurisdiction until released from its assignment. IST personnel will work with the local Incident Commander to ensure that command objectives are clearly communicated to US&R Task Forces. Control then reverts to FEMA for reassignment to another locality or demobilization home.

The central point of supervision and coordination of a Task Force is the Task Force Leader. Generally this position is held by a Deputy Chief or Assistant Chief from the sponsoring department. The Task Force Leader must meld the separate

disciplines of search, rescue, medical, and technical personnel into an integrated response unit. The Task Force Leader must also coordinate logistics with the IST and implement strategic and tactical assignments. In cases where multiple Task Forces are employed, overall coordination of US&R operations in the field is the responsibility of the IST

## Resources Which a US&R Task Force Needs from Receiving Jurisdiction

A FEMA US&R Task Force has been developed to be as self-sufficient as possible in an effort to minimize its impact on the locality requesting assistance. A jurisdiction impacted by a disaster does not need the additional responsibilities of having to provide food services, billeting, and equipment for incoming Task Forces.

The Task Forces cannot operate in a vacuum, however. The local jurisdiction will need to provide the following:

- *Clear identification of the receiving jurisdiction's command structure. One of the most important pieces of information that an incoming Task Force Leader will want to know is to whom he is to directly report (i.e., Incident Commander, Section, Branch or Division Officer, Sector Officer).*

- *Clear identification of a tactical assignment and location. This should include any size up or damage information, types of structure(s) involved, knowledge of victim entrapment/ locations, etc. It would be most beneficial for the locality to provide the Task Force with local maps of the area as well as pertinent building plans and/or structural drawings.*

- *Coordination assistance from local utility companies and the local medical director*

## Other Considerations When Receiving a US&R Task Force

There are some financial considerations that a locality should plan for in their disaster preparations.

An affected locality, once aware that the scope of the event is beyond their capabilities, requests assistance from the state. Subsequently, once the state is aware that the requirements of the event are also beyond its capabilities, a request for assistance is made to the Federal government. By law, any request for Federal assistance must be made by the affected state. Under the provisions of the

Stafford Act, once the request has been approved by the President, response and recovery costs are split between the Federal government and the requesting state. Currently, the ratio of funding responsibility is 75%/25%, Federal to state. The President may decide to change the ratio to 90%/10%, or in certain cases, the Federal government may provide 100% funding.

Currently a locality that requests a US&R Task Force may not have to pay any costs for its services. The affected state is usually responsible for its share. This should be determined in conjunction with the appropriate state emergency management agency. Transportation for a Task Force, both en route and return home, is handled by FEMA, usually in conjunction with the U.S. Air Force. The home jurisdiction of a responding Task Force bills all costs for the deployment of their Task Force directly to FEMA for reimbursement. Therefore, a requesting locality does not have to provide initial funding nor request reimbursement from the Federal government for US&R Task Force assistance.

# CHAPTER 12: PROFILES OF TECHNICAL RESCUE TEAMS

**T**he following section provides some insight into the structure and organization of several existing rescue teams across the country. Some of these teams are well established; others are in their infancy. They serve as examples as to how different organizations recognized a need for specialized rescue, established the necessary support systems, and overcame the many diverse challenges that all teams must face before they are ready to respond to technical rescue incidents.

Some of the teams discussed here were formed from within existing fire departments that wanted to enhance their rescue capabilities in specific areas. Other teams utilize personnel and equipment from several different departments to form consolidated teams. Some of the teams are not fire department based, but have found a particular niche for their technical rescue capabilities.

## FAIRFAX COUNTY (VA) FIRE AND RESCUE DEPARTMENT

The Fairfax County Fire & Rescue Department was one of the first departments to institute a formal technical rescue team in the nation. The combination department staffs the team with both career and volunteer personnel. The team began as a single discipline trench rescue team in the early 1980s when the department was experiencing the occurrence of trench collapse incidents which it was not prepared to handle. This was due to the tremendous increase in commercial and residential construction in the building boom at that time.

In 1982, the department conducted rescue operations at a fatal trench collapse that received wide local media coverage. The notoriety of this incident, coupled with a request to the County Board of Supervisors, resulted in the initial appropriation of $90,000 to address training and equipment procurement for trench rescue operations. The first trench rescue class was conducted in 1983 for both career and volunteer personnel.

Shoreform sheeting panels, oak shoring, pneumatic jacks, and other appropriate equipment was purchased for the team. The department identified two of the 30 fire stations as having responsibility for trench rescue operations. Personnel at these stations staff two of the County's seven heavy rescue squads assuming primary responsibility for these operations. In addition, a joint partnership was explored with a local trade group, the Heavy Construction Contractors' Association (HCCA), which provided funding for a tow vehicle for one of the two l&foot enclosed trailers used to store and transport the majority of the trench rescue equipment.

Over the next several years, the emerging team began to handle not only trench rescue operations but also other functions. As such, the scope of the trench rescue team was broadened to include other tactical responsibilities. In 1986, the team's responsibilities were formally broadened to include structural collapse, confined space, high angle/advanced rope operations and bus/metrorail extrications. The team's name was changed to the Technical Rescue Team to reflect its new capabilities.

In addition, a third station staffed with certified personnel was added to the program. Minimum staffing requirements were mandated for the three heavy rescue squads at the "Tech Rescue" stations, requiring all personnel to be Tech Rescue certified. Dispatch algorithms were developed for various technical rescue-type incidents. The standard dispatch, along with the usual complement of engines, trucks, and BLS/ALS units is the two closest Tech Rescue Squads and one of the two tow unit/equipment trailers for any technical rescue incident. With minimum staffing requiring three Tech Rescue certified personnel on each squad and two on the tow unit/trailer, a minimum of at least eight certified personnel are assured immediately on any technical rescue incident.

Initial training for new team members consists of four modules spanning a total of ten days (80 hours) as shown in Figure 7-2. Full team training for on-duty members is conducted on a recurring basis every other month. Tech Rescue units are placed out of service and conduct classroom and practical training at the Fire & Rescue Training Academy. This recurring bi-monthly full scale training is alternated with shorter two-hour practical sessions. A training officer is sent to the three Tech Rescue stations on the alternate months to conduct this training. All training is replicated on each of the three shifts.

In 1987, the Fairfax County and Metro-Dade (FL) technical rescue teams were approached by the Federal Office of Foreign Disaster Assistance to function as its international disaster response teams. In 1990, the department became involved in the initial development of FEMA's Urban Search and Rescue Response System for domestic disaster response.

The complement of three fully staffed and equipped Tech Rescue heavy rescue squads and two identically equipped equipment trailers is backed up by a comprehensive disaster response cache valued at approximately $600,000. The disaster cache is available for local, national or international response. The 110 technical rescue response personnel are composed primarily of career and volunteer members and augmented with non-department specialist volunteers (i.e., canine handlers, structural engineers, emergency trauma physicians, etc.).

## LOS ANGELES COUNTY (CA) FIRE DEPARTMENT

The Los Angeles County Fire Department's technical rescue team, termed the **Urban Search and Rescue (US&R) Team,** is designed around career staffing of nine truck companies with certified rescue personnel. It works in concert with the department's Air Operations (helicopter) Unit. In addition, a primary response unit, US&R 1, responds countywide on all technical rescue incidents. This complement is augmented with nine support trailers carrying additional shoring, generators, lighting and support equipment. The team is responsible for trench collapse, structural collapse, aircraft crashes, bus/metro rail incidents, swift water rescue, confined space rescue, and helicopter operations.

Los Angeles County maintains three levels of technical rescue training. All department personnel are trained to Level I (awareness level). The certified personnel staffing the nine US&R truck companies are trained to Level II (specialist), while the personnel staffing the primary US&R 1 unit are Level III (instructor level). The department currently maintains approximately 350 Level II personnel backed up by 60 Level III. Personnel expressing interest in joining the team must have attained prior certification on their own in state-certified Rescue Systems I and II, Swift Water Rescue I and II, and a 16-hour trench rescue class.

Continuing education training is scheduled on a two-year cycle. Level II personnel are provided training on a quarterly basis throughout this cycle. An 8-hour training session per quarter is conducted with personnel attending off duty. Eight separate quarterly sessions are conducted over the course of the two-year schedule with confined space and water rescue modules being given once each year (twice per two-year cycle). Level III personnel attend 8-hour sessions off duty on a monthly basis.

Los Angeles County Fire Department also sponsors a full FEMA US&R Task Force which has developed and procured a full response equipment cache and formally operated during the Northridge earthquake in Southern California in January 1994 and the Oklahoma City bombing in April 1995.

## MONTGOMERY COUNTY (MD) DEPARTMENT OF FIRE AND RESCUE SERVICES

The fire and rescue departments of Montgomery County, Maryland, provide technical rescue services through a combination of career and volunteer personnel from 19 fire departments and rescue squads from 34 stations. The county Department of Fire and Rescue Services coordinates several different technical rescue teams.

The Montgomery County Department of Fire and Rescue Services' Collapse Rescue Team is a combination career and volunteer technical rescue team sponsored by the Rockville Volunteer Fire Department. The team is responsible for trench rescue, structural collapse, confined space and mass casualty incidents. The program is managed by a 10-person management core group, and the department currently maintains approximately 120 certified personnel. The collapse rescue team is part of the county's FEMA US&R Task Force.

Initial certification training for new members consists of students having prerequisite attendance at the county fire-rescue training academy's 60-hour Practical Rescue training course. This is followed by specialized team training: 40 hours of trench rescue training, 40 hours of structural collapse training, and 24 hours of confined space training. Recurring training consists of weekly training sessions conducted every Tuesday for all on-duty certified personnel. As such, each 6-hour session is presented three times to cover all three shifts.

The team has equipment segregated by function which is stored and transported on a platform on demand (POD) unit. Structural collapse, trench rescue, and mass casualty comprise the three equipment PODs. The department also maintains two 40-foot trailers for disaster response US&R missions. The units are placed at several different fire stations, where on-duty trained personnel can respond as necessary. Montgomery County dispatches an initial assignment for a confirmed structural collapse or trench incident of one ALS unit, one engine company, one truck company, two heavy rescue companies, a duty officer, and the Collapse Rescue Team. In addition, the appropriate equipment POD (depending upon incident type) and a technical rescue support unit (incorporating a small crane, shoring lumber, air compressor, etc.) is dispatched. Individual team personnel are contacted and dispatched by personal pager and respond from their assigned stations or from off duty. As such, an initial response for a technical rescue incident results in a minimum of 12 certified collapse rescue personnel, with additional rescuers available for assignment.

Maryland Task Force 1 responded to a mine collapse in Norton, Virginia, in 1992, and has been mobilized for mission response for Hurricane Emily in August 1993 and the Northridge earthquake in Southern California in January 1994. The team also responded to the Oklahoma City bombing in April 1995.

A separate team exists for technical rope rescue operations. The Montgomery County Department of Fire and Rescue Services' Special Evacuation Tactics (SET) Team specializes in low-angle, high-angle, and urban rope rescue, and patient evacuation. The combination career and volunteer team members must complete the 60 hour practical rescue curriculum covering the basics of rope, trench, confined space, swift water, and extrication. Members then receive additional specialized training focusing on rope rescue and rope evacuation techniques. The SET Team has one weekend training drill per month. All members are EMTs or Paramedics; all the paid personnel and most of the volunteers are cross-trained as firefighters.

The team is decentralized, with personnel responding from on or off duty via dispatched paging system. The team has performed several rescues the last few years. Many of their high angle responses are in an urban environment. Team members bring their own cache of personal gear with them; the county will dispatch a heavy rescue squad to provide additional personnel and equipment. The county is currently evaluating whether to assign the SET Team to a single station.

The Montgomery County Department of Fire and Rescue Services' Underwater Rescue Team (URT) performs dive rescue and recovery operations. The team is open to any career or volunteer member that has SCUBA training. Team members are then trained to national standards for dive rescue. The team was initially decentralized, like the SET Team, but has recently become sponsored by the Germantown Volunteer Fire Department where their equipment is now based on a ladder company and inflatable rescue boat.

A fourth specialized service is water rescue. Several Montgomery County fire departments have rescue boats and trained personnel. Specialized swiftwater rescue is provided by the Cabin John Park Volunteer Fire Department which provides swiftwater and ice rescue to the Potomac River. The department has been providing water rescue to this area for several decades. In the last 10 years the Cabin John Park water rescue team, the "River RATS" (River Rescue And Tactical Services), has evolved into a well trained and equipped unit consisting of approximately 40 personnel made up from paid and volunteer members at the department. The team responds from two fire stations. Funding is provided by a combination of county tax funds, state grants, and donations.

Team members are trained in emergency care (EMT or paramedic), technical rope rescue, ice rescue, water rescue, and boat operations from local rescue and privately-provided programs. Members receive progressive levels of training as boat crews, boat operators, and airboat pilots. The department has three inflatable rescue boats, one airboat, two aluminum boats, and two four-by-four boat support unit vehicles, which hold water rescue and high angle rescue equipment, as well as two all terrain bicycles for response along the many bike and hiking trails along the river.

Montgomery County recently constructed a collapse rescue training building at their fire and rescue academy. The site consists of several collapse rescue operations and a mock collapsed building known as the "Rescue Mall." This site has been used for FEMA US&R Rescue Specialist training by several FEMA US&R teams.

## CHATTANOOGA-HAMILTON (TN) EMERGENCY SERVICES

The Chattanooga-Hamilton Cave and Cliff Rescue Unit is part of the Chattanooga-Hamilton County Emergency Services. The team provides specialized rescue for cave and wilderness high angle incidents.

Established in the 1950s to provide cave rescue in one of the world's premier recreational spelunking areas, the team has evolved into a specialized unit that has been called for international response. The team consists of 18 volunteer members. All members have had at least five years of caving experience and have been recommended for the team by the National Speleology Society and a local caving club. The team is supported by county tax funds and donations. Team members are alerted by a paging system and respond in several four-by-four utility vehicles. Members of the team are minimally certified to Cave Rescuer 1 and first responder, though many members have advanced training. Some team members are medical doctors.

## CITY OF JACKSON (MS) FIRE DEPARTMENT

The Jackson Fire Department Underwater Rescue Team performs dive rescue in the City of Jackson and surrounding jurisdictions. The team consists of 30 members of the municipal fire department that respond from their assigned stations in the event of a water rescue.

The team is supported by fire department tax funds. Members are trained and certified according to national diving standards. The team has one response vehicle and several boats.

## VIRGINIA BEACH (VA) FIRE DEPARTMENT

The Virginia Beach Fire Department maintains a career technical rescue team that is also a component of an overall regional team in the Virginia Tidewater area. The response team is centered around staffing five truck companies with certified personnel backed up by a primary response unit, Tech 1. In addition, a utility support vehicle is available as well as a 24-foot trailer carrying trench rescue equipment and a 40-foot tractor/trailer carrying structural collapse equipment.

The team is part of the Tidewater Regional Heavy and Tactical Rescue Team, consisting of rescuers from Virginia Beach, Norfolk, Norfolk Naval Base, Chesapeake, Little Creek, Franklin, and Newport News. The Virginia Beach Fire Department, in conjunction with the Tidewater Regional Team, sponsors a full FEMA US&R Task Force. The team conducted Task Force operations after the Petersburg, Virginia tornadoes in September 1993 and the Oklahoma City bombing in April 1995.

Initial training for certified personnel consists of a total of 17 days training in the following modules: structural collapse, confined space rescue, helicopter operations, and advanced rope rescue. The team currently maintains approximately 100 certified personnel backed up by awareness training for all department personnel. Virginia Beach conducts recurring team training on a quarterly basis and participates in quarterly regional training sessions. As such, each team member usually receives 16 hours of training per quarter.

## BALTIMORE COUNTY (MD) FIRE DEPARTMENT

The Baltimore County Fire Department's Advanced Technical Rescue Team (ATRT) was started in the early 1980s in an effort to establish a high-rise emergency aerial team in cooperation with the Maryland State Police Aviation Division. As technical rescue developed, the team incorporated trench rescue, water rescue, and confined space rescue. The team grew with the assistance of several grants, including one from Nations Bank for two search and rescue canines. A 22-foot truck and over $70,000 worth of equipment was also donated. The team has two vehicles located at a single station.

The team consists primarily of members of the Baltimore County Fire Department, supplemented by volunteers from fire companies within Baltimore County. Each shift has 16 ATR Team members at separate stations. The on-duty team members respond when dispatched in staff vehicles from their normal stations; off duty personnel respond via pager. The team conducts monthly training and responds to approximately two dozen calls per year.

The team conducts an annual trench rescue training weekend attended by 250 rescuers. The ATR Team has also developed a technical rescue awareness video for all members of the Baltimore County Fire Service. The team has a mutual aid agreement with five other jurisdictions and can be flown to rescue locations by the Maryland State Police.

## FAIRFIELD COUNTY (CT) TECHNICAL RESCUE TEAM

The Fairfield County Technical Rescue Team provides an example of a team in its infancy. It does not yet have technical rescue response capability. Nevertheless, the team provides an example of a group of rescuers that have identified a need in their area and are attempting to address that need.

The goal of the team members is to provide trench, collapse, confined space, rope, dive, and water rescue in Fairfield County, Connecticut. The members of the team are volunteers from the 20 paid and volunteer fire departments in the county. The team is currently funded by providing confined space and rescue training to local groups, and it is arranging to have a vehicle donated. The team has had difficulty enlisting the support of many of the independent fire departments within the county.

Members of the team have pursued training through various agencies, including from state rescue programs. Team members are continuing to push for increased acceptance on the part of the fire departments, with the goal of becoming a consolidated, countywide team for technical rescue operations and training.

## ST. CHARLES (MO) FIRE DEPARTMENT

The St. Charles Fire Department Technical Rescue Team was recently formed to perform high angle, confined-space, collapse and trench rescue. The team consists of 24 paid members from the St. Charles Fire Department. It is supported by tax funds and donated money and equipment.

The Department is slowly phasing in its technical rescue abilities. Members are currently being trained in rope rescue techniques. The department is training with other City agencies such as the public works department so that its can be called in to support on technical rescue incidents. The goal of the Department is to be capable of performing all four types of technical rescue within two years. The team will be available for local and regional service upon request. The fire department has purchased a rescue-engine for the station that will eventually house the team.

St. Charles started their team after realizing that the citizens of the community expected them to have the ability to perform these types of rescues. The team hopes that additional funding may be provided by becoming the emergency response team for several local industrial plants.

## INDIANAPOLIS (IN) FIRE DEPARTMENT

The City of Indianapolis Fire Department has developed a special program for training personnel in technical rescue operations. The Basic Emergency Rescue Technician (BERT) program trains Indianapolis firefighters in rescue disciplines including collapse, trench, and confined space rescue, vehicle and machinery extrication, rope and swiftwater rescue techniques, and search operations. The program emphasizes a team oriented rescue philosophy over specific rescue techniques. It includes information on safety, command, physics, and rescue management.

The BERT program is about 200 hours in length and is divided in to eight specific modules. The curriculum was developed by reviewing the types of urban rescue operations that Indianapolis firefighters had been called to perform over a 10 year period. The training program is structured loosely on the design of hazardous materials training, with personnel trained to awareness, operations, technician, and specialist levels. The BERT training was started in 1988; there are currently over 200 BERT certified members of the department.

The Indianapolis Fire Department has six stations designated as technical rescue "Task Forces." Each station specializes in two areas of technical rescue so that personnel can maintain high levels of training in certain disciplines. Personnel respond together in a single rescue vehicle for technical rescue incidents.

## PHILADELPHIA (PA) FIRE DEPARTMENT

The Philadelphia Fire Department rescue company, Heavy Rescue 1, was established in the late 1980s in an effort to provide specialized heavy and technical rescue service. The department developed a proposal for establishing a heavy rescue company and surveyed other departments that had heavy rescue companies. As the City of Philadelphia was facing budget shortfalls and could not fund the team out of budget money, the members of the department successfully sought alternative funding sources. They solicited private businesses in the community to sponsor the team. Businesses were presented with a list of needed equipment and asked to purchase a single item as a team sponsor. Over $100,000 in equipment was purchased by local businesses.

The team refurbished a beverage truck donated by the Coca-Cola Company into a equipment vehicle, and rehabilitated an old engine by removing the pump and tank to carry equipment for collapse rescues. Members were recruited from the department, and emphasis was placed on special experience in skills such as electrical or carpentry trades. The heavy rescue company was placed in service in 1991. In 1994, the department placed in service a new heavy rescue truck to replace the old beverage truck.

The rescue company conducts specialized training with other city agencies such as the electric and gas company. Heavy Rescue 1 is dispatched on all working box alarms and multiple alarm fires, special rescue incidents such as collapses and confined space rescues, extrications, and water rescues.

## CHESTERFIELD (VA) FIRE DEPARTMENT

The Chesterfield Fire Department provides technical rescue services to a suburban area outside of Richmond, Virginia. It has developed a departmentwide approach to technical rescue training. Several years ago Chesterfield set out to train personnel to an awareness level for technical rescue operations. Over the course of two years, all 350 career firefighters and about 25 percent of the volunteer firefighters received awareness level training. These personnel are trained to recognize technical rescue hazards and to conduct initial size-up and stabilization procedures. Some technical rescues can be handled at this stage of response, which usually consists of an engine company. Awareness training covers rope rescue, vehicle and machinery extrication, confined space operations, trench rescue, urban search and rescue, and water/ice rescue. The training was conducted in five hour modules over several months to career personnel. A course for volunteers is offered over several weeks each year. The Chesterfield Fire Department has also begun training personnel to an operations or Level II training. These personnel receive 80 hours of training in rope, vehicle, confined space, trench, and heavy equipment rescue.

The department plans to establish a level III training program where personnel will receive intensive training in a particular area of technical rescue. A fourth level for technical rescue training which will cover technical rescue incident command is also being established. The incident command class will be available for any personnel and will focus on special com-mand considerations for technical rescue incidents. Each shift has instructors for technical rescue. The special operations personnel are supported by specialized hazmat team and SCUBA Rescue (dive) team.

## ALLEGHENY COUNTY (PA) DELTA TEAM

The County of Allegheny has a unique resource available for heavy rescue operations, the

Delta Team. This rescue team is part of the Allegheny County Department of Special Services and Maintenance Operations. The department set about to form their own heavy rescue group after several requests for assistance from fire and rescue agencies for their specialized heavy equipment. Realizing that their personnel had the experience of using heavy machinery, but lacked the training for rescue operations, the department began an intensive process of training its personnel and funding additional equipment. It provides confined space, collapse, trench, and water rescue assistance.

The personnel on the team include emergency managers, heavy equipment operators, ironworkers, riggers, plumbers, carpenters, electricians, engineers, repairmen, and mechanics. The department has over 400 public works vehicles at its disposal, including cranes, bulldozers, all terrain vehicles, trailers, and buses. Police and fire radios are installed in many vehicles to facilitate emergency communications. The department also has converted several trailers for use during long term operations. The trailers are set up as showers, toilet facilities, sleeping quarters, a cafeteria, and a field hospital. The department has a plethora of shoring equipment at its disposal.

A marine division within the team has several boats at its disposal for flood situations, from small inflatable rubber boats to a 28-foot inboard diesel powered boat. The team also has equipment for mitigating hazardous materials incidents, including booms and absorption materials, and entry suits with SCBA.

Team members train on a monthly basis and have a several day disaster training drill once a year.

## HYANNIS (MA) FIRE DEPARTMENT

In addition to providing fire, rescue and EMS services to Cape Cod, the Hyannis Fire Department has established a technical rescue training program for all members of the department. Operating out of one of the busiest stations in Massachusetts, the team responds with a heavy rescue company that carries most of its technical rescue equipment.

All 48 fire department personnel are trained in confined space rescue, and all are currently receiving training in rope and trench rescue. In

addition, the department has an eight member dive rescue team. Rope training is divided into a Rope I and a Rope II course, and trench rescue training is modeled after the Virginia Beach, Virginia trench rescue program. All personnel will eventually be trained for rope and trench rescue. Some members of the department are part of the Massachusetts FEMA US&R Task Force Team.

## HAILEY VOLUNTEER FIRE DEPARTMENT AND WOOD RIVER (ID) FIRE PROTECTION DISTRICT

Members of the Hailey Volunteer Fire Department joined with the members of the Wood River Fire Protection District last year to begin forming a technical rescue training program between the two departments. Covering a small town and large rural area, the departments realized that they were the only ones available to provide any sort of technical rescue. Department leaders formed a three tiered training plan that would take over six years to train their personnel to the ability they desired. The first tier of training has been devoted to developing a basic awareness and technical proficiency in technical rescue. So far, team members have received training as swiftwater rescue technicians, and to a first responder level for rope rescue and confined space rescue. Training in heavy extrication and collapse rescue is planned for the future. Some team members have been sent to several outside technical rescue training programs to receive initial and advanced training.

Both departments have been buying their equipment gradually, setting aside several thousand dollars each year as their members receive more training and experience. They currently have built a cache of rope, water, and confined space rescue equipment. Technical rescue training has also been incorporated into the Hailey Volunteer Fire Department weekly training sessions for their volunteers. There have been some minor difficulties in coordination between the two departments, but both organizations support the continued development of their technical rescue capability.

## TAMPA (FL) FIRE DEPARTMENT

The Tampa Fire Department provides rope rescue, confined space rescue, and water rescue services. Under a special operations division, 60 personnel provide technical rescue for rope and confined space incidents from an engine and ladder company. Other components of the special operations division include hazmat, marine firefighting, and extrication. A separate Tactical Medical Response Team provides wildland search and rescue and water rescue service under the Tampa Fire Department's Rescue (EMS) Division.

The department's technical rescue services are currently evolving. There is currently no Florida state certification for rescue, so the department is establishing its own standards based on technical rescue programs in the state and around the country.

All Tampa Fire Department personnel receive a basic awareness level of training for some aspects of technical rescue, such as confined space rescue.

## MILWAUKEE (WI) FIRE DEPARTMENT

The City of Milwaukee's Heavy Urban Rescue Team is responsible for trench, below grade, collapse, confined space, and rope rescue. The team also incorporates much of the capabilities of Milwaukee's tunnel rescue team, which was placed out of service last year. The Heavy Urban Rescue Team has approximately 100 members that have received training in technical rescue. Initial training is conducted through a vocational-tech program in

Cleveland, Wisconsin, with continued training on a monthly basis within the department. The department has brought instructors from various parts of the country to present technical rescue training programs.

The team is based on an engine company and a ladder company providing 10 personnel per shift available for initial response on technical rescues. Additional personnel may be called back from off duty or from other station assignments if necessary. The team is currently attempting to procure a tractor trailer type vehicle to carry heavy rescue equipment.

Funding for the Heavy Urban Rescue Team is provided though the City of Milwaukee Fire Department's budget and through a special annual private grant of $50,000. The team is available for mutual aid upon request. The specialty team charges $500 dollars for the first two hours of service, and $250 per hour thereafter.

The Milwaukee Fire Department also has a Dive Rescue team based out of a single company. There are 35 members of the team, with five assigned to the dive team each shift. Divers are trained to the PAD1 standards for dive rescue and many have advanced life support certifications.

# APPENDIX A: SAMPLE MUTUAL AID AGREEMENTS

## Palm Beach County, Florida
## State of California

FIRE-RESCUE

INTERLOCAL EMERGENCY SERVICES AGREEMENT

FOR

GOVERNMENTAL AGENCIES IN PALM BEACH COUNTY

APRIL 15, 1994

# INTERLOCAL EMERGENCY SERVICES AGREEMENT

## Table of Contents

# INTERLOCAL EMERGENCY SERVICES AGREEMENT

THIS INTERLOCAL AGREEMENT for emergency aid and assistance is made and entered into in Palm Beach County, Florida, as of this__day of _____ 199_pursuant to 'the provisions of Section 163.01, Florida Statutes, the "Florida Inter-local Cooperation Act of 1969" as amended, by and between the various parties executing this agreement, each one constituting a public agency as defined in Part I of Chapter 163, Florida Statutes.

## W I T N E S S E T H :

WHEREAS, Part I of Chapter 163, Florida Statutes, permits public agencies as defined therein to enter into inter-local agreements with each other to jointly exercise any power, privilege, or authority which such agencies shall have in common and which each might exercise separately; and;

WHEREAS, it is the design, purpose and intention of the parties hereto to permit said parties, individually and collectively, to make the most efficient use of their respective powers, resources and capabilities by cooperating in the use of their respective powers, resources and capabilities in regard to tire, heavy rescue, emergency medical services, and related logistical, strategic and administrative support, communications services, weather emergencies and disaster relief functions and, on a basis of mutual advantage, to provide services and facilities in a manner most consistent with the geographic, economic, demographic and other factors influencing their respective needs and the development of their respective and joint communities; and

WHEREAS, each party hereto maintains a Fire Department, Fire-Rescue Department or Public Safety Department with trained emergency service personnel and related apparatus and equipment; and

WHEREAS, at times, one of the parties hereto- may have fire fighting, rescue, emergency medical service, disaster relief and related demands made upon its equipment or personnel, or

4

INTERLOCAL EMERGENCY SERVICES AGREEMENT

both, greater than the capacity of the equipment or personnel available within its own jurisdiction; and

WHEREAS, during those events which cause demands greater than the capacity of the equipment or personnel resources available to a party hereto, it would be advantageous to that party to have available to it the equipment or personnel, or both, of one or more of the other parties for backup purposes, direct assignment to an active incident, or the management of a disaster; and

WHEREAS, the parties hereto acknowledge that said emergency events and disasters occur without prior warning, without a set pattern or frequency and without regard to life, limb or property; and

WHEREAS, the parties hereto further recognize that there is a great mutual advantage in providing, prior to any emergency operation or disaster, for mutual and/or automatic ,aid and assistance, planning, deployment analysis and projections, mutual backup .and cooperative use of the resources available among the affected parties, in order that lives and property may be saved; and

WHEREAS, said mutual and/or automatic aid, and other cooperative use of resources benefits all directly or indirectly concerned; and

WHEREAS, it is the intent of the parties to this Agreement to provide for mutual aid in general, automatic aid in specific instances when agreed to between the participating parties, and special operations when authorized by supplemental agreements, to improve efficiency or for unforeseen emergencies beyond the normal capabilities of a single party.

# INTERLOCAL EMERGENCY SERVICES AGREEMENT

WHEREAS, <u>It is not the intent of the parties that the mutual aid aspect of this agreement subsidize normal day-to-day operations of another participating party;</u> and

WHEREAS, participation in this agreement shall not diminish any existing local government's process or power;

NOW, THEREFORE, in consideration of the premises and mutual covenants and promises contained herein, and other good and valuable consideration, the receipt of which and the adequacy of which are mutually acknowledged, with all parties accordingly waiving any challenge to the sufficiency of such consideration, it is mutually covenanted, promised and agreed by the parties hereto as follows:

1. <u>AUTHORITY: GENERAL RESPONSIBILITIES: CONDITIONS PRECEDENT:</u> This Inter-local Agreement is entered into pursuant to the provisions of Section 163.01, Florida Statutes, commonly known as the "Florida Inter-local Cooperation Act of 1969", and all provisions of said act are made a part hereof and incorporated as if set forth at length herein, including, but not limited to, the following specific provisions; as well as the apparatus, staffing, dispatch, response and communication system standards as hereinafter enumerated:

(a)     All of the privileges and immunities and limitations from liability, exemptions from laws, ordinances and rules, and all pensions and relief, disability, workers' compensation and other benefits which apply to the activity of officers, agents or employees of the parties hereto when performing their respective functions within their respective territorial limits for their respective agencies, shall apply to the same degree and extent to the performance of such functions and duties of such officers, agents or employees extraterritorially under the provisions of this Inter-local Agreement.

*(b)*    This Inter-local Agreement does not and shall not be deemed to relieve any of the patties hereto of any of their respective obligations or responsibilities imposed upon them by law, except to the extent of the actual and timely performance of those obligations or responsibilities by one or more of the parties to this Agreement, in which case performance provided hereunder may be offered in satisfaction of the obligation or responsibility.

*(c)*    Nothing contained herein shall be deemed to authorize the delegation of the constitutional or statutory duties of the State, County, or Municipal Officers.

*(d)*    As a further condition, the following minimum apparatus, personnel and response standards must be met by all parties to this agreement:

     *(1)*    Apparatus - At least one, triple combination pumper with a capacity of at least 1,000 G.P.M., with 400 gallon minimum tank size, and 1,200 feet of 2 1/2 or larger fire hose and other equipment specified in Paragraph 3, shall be provided by every party hereto.

     *(2)*    Staffing - As specified in Paragraph 6.

     *(3)*    Dispatch time frame - Dispatch of apparatus and personnel will be completed and confirmed within five (5) minutes of receipt of request for assistance.

     *(4)*    Response time frame - Apparatus and personnel will be enroute within two (2) minutes of being dispatched following receipt of request for assistance.

     *(5)*    Communication system access - As specified in Paragraph 7.

## 2. DEFINITIONS

ADVANCED LIFE SUPPORT (A.L.S.) shall mean emergency medical services as defined by 401.23, Florida Statutes.

AUTOMATIC AID shall mean supplementary assistance provided to one or more parties to this Agreement by another party by separate Agreement or Resolution.

BASIC LIFE SUPPORT (B.L.S.) shall mean emergency medical services as defined by 401.23, Florida Statues.

EMERGENCY AID AND ASSISTANCE shall mean the full scope of fire and emergency medical services, except those exempted by this Agreement.

EMERGENCY MEDICAL SERVICES (E.M.S.) means the activities or services to prevent or treat a sudden critical illness or injury and to promote pre-hospital emergency medical care for the sick and injured.

EMERGENCY MEDICAL TECHNICIAN means a person providing emergency medical services as defined by 401.23, Florida Statutes.

ENGINE/PUMPER shall mean a pumping apparatus meeting the requirements of National Fire Protection Association Standard 1901.

FIRE CHIEF shall also mean "Fire Administrator, Fire Director, Director of Emergency Services or Public Safety Director.

FIRE FIGHTER shall mean a person who is certified by the State of Florida under Florida Statute, Section .633.35, as amended, .and possessing. either. a valid Certificate of Tenure, Compliance or Completion of FireFighting Minimum Standards.

INCIDENT COMMAND SYSTEM shall mean the procedures adopted by the Fire Chiefs Association of Palm Beach County for emergency incident management.

INCIDENT COMMANDER shall mean the person at the scene of an emergency incident who is designated to command the incident.

MUTUAL AID shall mean mutual and reciprocal aid and assistance between two or more parties to this agreement.

PARAMEDIC means a person providing emergency medical services as defined by 401.23, Florida Statutes.

PROTOCOLS shall mean the standing orders and procedures utilized by E.M.S. personnel to render both A.L.S. and B.L.S. care and treatment.

QUINT shall mean a pumper/aerial apparatus meeting the requirements of both National Fire Protection Association Standards 1901 and 1904.

SPECIAL OPERATIONS shall mean response to, and mitigation/management of hazards relating to hazardous materials incidents, high angle rescue, confined space rescue, and water rescue operations.

TRUCK/AERIAL shall mean an aerial ladder apparatus meeting the requirements of National Fire Protection Association Standard 1904.

3. APPARATUS, EQUIPMENT AND ACCESSORIES: It is agreed that at the time of the execution hereof each party represents that it' has the apparatus listed in Appendix A, and that said apparatus is in good working order and condition. Only the equipment listed in Appendix A, as supplemented pursuant to paragraph 4 of this Agreement, shall be subject to this Agreement. In addition to the specifically described major pieces of apparatus listed, said apparatus shall be

deemed to include all normal appliances, tools, accessories and portable equipment associated therewith and normally contained thereon as recommended by the basic required equipment described in the latest editions of National Fire Protection Association Standards No. 1901 and/or 1904, as applicable, and the required equipment specified by the State of Florida, Department of Health and Rehabilitative Services, Division of Emergency Medical Services, in Chapter 10D-66, Florida Administrative Code, for Advanced and Basic Life Support Units, as amended from time to time, as applicable.

4. NEW APPARATUS, EQUIPMENT AND ACCESSORIES ACQUISITION: Upon the acquisition. of new apparatus, said apparatus, upon being placed in service, shall, without formal amendment hereto, be deemed included in Appendix A. The party acquiring such new apparatus shall notify the Administrative Board, in writing, of the acquisition of new apparatus as soon as the equipment is placed in service and deployed. The Administrative Board shall subsequently advise all parties to this agreement, in writing, of the acquisition of the new apparatus by publishing and distributing a revision to Appendix A.

5. CHANGE IN OR DELETION OF APPARATUS, EQUIPMENT OR ACCESSORIES: Nothing herein shall prohibit, any party from the free disposal or modification of any of its apparatus, equipment or accessories, or from temporarily placing all or part or a major part of its apparatus out of service for the purpose of training, maintenance or repair of the same. However, should any party actually or effectively dispose of, or eliminate fifty percent (50%) or more of its apparatus without immediate replacement, said disposal shall be reported immediately to the Administrative Board and, then, at the option of the other parties hereto, as determined by a majority vote of the Administrative Board, this Agreement may be conditionally terminated. A conditional termination shall take effect only if the party whose participation is being terminated fails to bring its designated in service, active equipment up to the level specified in the notice of conditional termination within the time limit set forth in the notice by the Administrative Board.

Such termination shall be in writing and shall not take effect until the Administrative Board has served written notice upon the City Manager, County Administrator, Chief Administrative Officer, Mayor or City/Town Clerk of the party whose participation is to be terminated.

6. <u>STAFFING:</u> All responding vehicles shall be staffed by the party responding to a mutual or automatic aid request. Staffing shall be a complement of a minimum of three (3)' Fire Fighters for each Engine and three (3) Fire Fighters per Truck, Aerial or Quint, and staffing as required by Chapter 401, Florida Statutes, as amended, for each emergency medical service unit, as well as appropriately trained personnel for support equipment as may be requested. It is understood that if additional staffing or personnel are on duty, called for duty, or available for the emergency duty in question, then said additional personnel shall be provided for this purpose.

7. <u>COMMUNICATIONS:</u> Recognizing that radio communication is necessary to successfully provide. fire fighting, emergency medical service, or disaster related assistance, the parties agree to provide the capability to access the following radio frequency(s):

F.C.C. Radio/Call Sign -    <u>KLL578</u>

<u>FREQUENCY(S)</u>          <u>DESIGNATION/USE</u>

154.265               Mutual Assistance

(a) This channel will be restricted to use by command personnel only as described in Paragraph 10.

(b)   Compliance with this section of the Agreement shall be the responsibility of each party within one (l) year of their execution of the Agreement.

8. RESPONDING PARTY/REQUESTING PARTY: "Responding Party" shall mean the party which shall furnish or be requested to furnish apparatus, equipment, and/or personnel, in response to the request of the party within whose jurisdiction the emergency necessitating such assistance occurs, which second party shall be known as the "Requesting Party",

9. RESPONSE TO REQUEST. FOR ASSISTANCE: The parties hereto, mutually agree to respond to mutual aid fire, emergency medical service, emergency or disaster calls, or requests of the other with their respective apparatus, equipment and associated personnel as herein above described, when requested to do so by the Requesting Party, subject to the terms,. conditions and understandings contained in this agreement and within the limits of good Fire Department practice and procedure, and further, the parties hereto, who have mutually agreed to assist each other by means of automatic aid and/or special operations responses in addition to mutual aid responses, agree to respond to the request for assistance with pre-determined apparatus, equipment and personnel within limits of Fire Department practice and procedure.

10.. OFFICIAL REQUEST: The following officials of the Requesting Party are authorized to request mutual and/or automatic assistance from the Responding party pursuant to this Agreement,

(a)    Fire Chief, Deputy or Assistant Fire Chief(s) or

(b)    The Senior Officer in Charge/Command of the Requesting Party's Fire Department, Fire-Rescue Department or Public Safety Department; or

(c)    the Incident Commander in charge of an incident in progress.

11. .REQUEST FOR ASSISTANCE: INFORMATION: The officer described in Paragraph 10 above requesting mutual or automatic aid shall give the following information at the time that the request for such assistance is made:

(a)  The general nature and type of emergency;

(b)  The location of the emergency and/or the Fire Station to be filled by responding resources;

(c)  The life or property hazard involved and the type of equipment and/or number of personnel requested;

(d)  Street routing information when necessary.

The initial request for assistance shall be transmitted by radio, telephone or other reliable method to the appropriate dispatch/communications center of the Responding Party.

12. JUSTIFIED FAILURE TO RESPOND: The parties recognize and agree that if for any reason beyond the control of the Responding Party the above-referenced equipment or personnel, or both, are not available to respond to a request for assistance within the limits of the Requesting Party, the Responding Party shall not be liable or responsible in any regard whatsoever for the Responding Party's failure to respond to such call. The reasons justifying a failure to respond shall include, but not be limited to the following:

(a) In the opinion of the Senior Officer in command of the Responding Party's Fire Department at the time of the request for assistance, the Responding Party would suffer undue jeopardy and be left inadequately-protected if the Responding Party. responds as requested;

(b) The requested equipment is inoperative at the time of the request for assistance; or

(c) The requested equipment is being utilized due to a previous emergency call.

13. INDEMNIFICATION

Each party shall bear its own responsibility and may only be liable for any claims, demands, suits, actions, damages and causes of action arising out of those actions resulting from that patty's travel to or from its own or a Requesting Patty's emergency or disaster site, or while deployed pursuant to this Agreement, and no indemnification or hold harmless agreement shall be in effect concerning such claims, demands, suits, actions, damages and causes of action.

14. DAMAGE TO EQUIPMENT: The Requesting Party shall replace, repair or reimburse the Responding Party for the actual cost of replacement or repair of any of the Responding Party's apparatus, tools, mechanical equipment or personal protective equipment which may be damaged or destroyed while at the Requesting Party's emergency or disaster site unless such damage or destruction is solely the result of errors, negligent acts or omissions of the Responding Party or any of its agents, employees or officials.

15. MATERIALS AND SUPPLIES: The Requesting Party shall, at the option of the Responding Party, either replace or reimburse the Responding Party for the actual cost of all materials, and supplies such as foam, dry chemicals, extinguishing agents, disposable protective garments and related equipment, medical supplies and consumables used for the benefit of patients at the emergency site or disaster, consumed, expended or utilized by the Responding Party in the course of rendering assistance pursuant to this Agreement while at the Requesting Party's emergency or disaster site.

16. CONTROL OF FIRE, RESCUE, EMERGENCY OR DISASTER SCENE: Once the Responding Party reaches the Requesting Party's emergency or disaster site, the Parties agree that the Requesting Party's Incident Commander shall direct the activities and deployment of personnel and equipment in the area where the emergency exists. Control of each Respective Party's personnel shall remain with each respective party as to the rendition of service, standards of performance, discipline of officers and employees and other matters incident to the performance of services by the Responding Party's personnel. The officer in command of the personnel of the Responding Party shall not be obligated to obey any order which said officer reasonably believes to be either in violation of the laws of the State of Florida, the United States of America, or any order which said officer believes will unnecessarily result in the likelihood of unreasonable risk of death or bodily injury to the agents, officers or employees of the Responding party, or in a loss of or damage to the Responding party's equipment. All parties to this agreement stipulate that they will utilize the "Incident Command System" adopted, and from time to time revised, by the Fire Chiefs Association of Palm Beach County for all appropriate operations covered by this agreement.

17. REQUESTS FOR ASSISTANCE LIMITED: A party may request assistance pursuant to this agreement only when the site of the emergency or disaster causing such a need is within the jurisdictional limits of the Requesting Party, or is within a jurisdiction which is served by a party to this Agreement pursuant to an interlocal service agreement or contract between two or more governmental agencies.

18. PRIORITY FOR SIMULTANEOUS CALLS: In the event of simultaneous or nearly simultaneous fire, rescue, emergency or disaster calls relating to emergencies located within both a Requesting and a Responding Party's boundaries, the call relating to the emergency located within the boundaries of a party shall take priority over the request for assistance from the Requesting Party.

19. PRIOR COMMITMENT OF EQUIPMENT: In the event that a Responding Party's equipment and personnel are already assigned or committed to an emergency located within Responding Parties limits, said equipment and personnel shall not be released to respond to the emergency call of the Requesting Party until such time as, in the sole and absolute discretion of the officer in charge of the Responding Patty's Fire Department, it is determined that the Responding Party's equipment and personnel can be released to respond to the Requesting Party's request for assistance. Only that portion of the Responding Party's equipment and personnel as the Officer in Charge of the Responding' Party's Fire Department shall deem available for release at that time shall be released to the Requesting Party's emergency or disaster site.

20. PRIORITY FOR SUBSEQUENT CALLS: In the event that the Responding Party's equipment and personnel are assigned to a location within the Requesting Party's Jurisdictional limits and an emergency call relating to a fire, rescue, emergency or disaster occurring within the Responding Party's boundaries is received, the Requesting Party shall, immediately upon being notified by the Responding Party's officer in command at the site of the Requesting Party's emergency of such circumstances, request such additional outside assistance from other parties to this agreement, as would timely and effectively permit the release of the Responding Party's equipment and personnel so as to enable same to timely respond to the Responding Party's emergency call site or, if same is not practical or feasible, the Requesting Party agrees that the Responding Party shall be permitted to immediately leave the emergency or disaster site within the territory of the Requesting Party and respond to the fire, rescue, emergency or disaster site within the Responding Party's territory. In any event, should the officer in command of the Responding Party's fire department require that its equipment and personnel return to an emergency site within its boundaries in response to a fire, rescue, emergency or disaster call, the parties agree that the Responding Party has the absolute right to immediately- return to the emergency site within its jurisdictional limits.

21. AMENDMENTS

(a) An amendment to this Agreement may be initiated by any party to the agreement, however, said amendment must be presented' to the Administrative Board for consideration, deliberation, determination and action and approval.

(b) All parties hereby agree to submit to their legislative bodies for final approval, additions to and deletions from this agreement which have been approved by a two-thirds vote of the Administrative Board.

(c) Approval of such proposed amendments by two-thirds of the participating agencies' legislative bodies shall make such amendments effective, and shall be binding upon all parties continuing to participate in this agreement.

(d) Amendments hereto shall be filed with the Clerk of the Circuit Court of Palm Beach County.

22. ADMINISTRATIVE BOARD:

(a) The Administrative Board shall consist of the Fire Chief from each of the participating jurisdictions. The Administrative Board shall have the authority to adopt rules of procedure for conducting meetings and for ruling upon disputes, disagreements, grievances and other matters to be determined by the Administrative Board pursuant to this Agreement.

(b) No party to this Agreement shall in any manner be obligated to pay any debts or liabilities arising as a result of any action of the Administrative Board.

(c) The Administrative Board members have no authority or power to obligate the parties in any manner, except that the Administrative Board shall have the power to suspend any party's further participation for non-compliance with any provisions of the Agreement.

23. TASK FORCES: The Administrative Board may organize various task forces as it deems appropriate. Task forces shall be assigned specific topics for research and development. Recommendations from task force(s) shall be forwarded to the Administrative Board upon completion of assigned work.

24. PRIORITY OBJECTIVES: In the interest of uniformity and cost effectiveness it is hereby understood and agreed that task forces will initially be formed by the Administrative Board to address the following topics: Incident Command; Training; Facility Location Planning; Communications; Joint Purchasing; and Equipment Specifications and Standardization,

25. DISPUTES, DISAGREEMENTS AND GRIEVANCES: Disputes, disagreements or grievances which cannot be voluntarily resolved by the parties directly involved shall be resolved by the Administrative Board by majority vote, whose decisions in such matters shall be final and binding upon all parties.

26. INTERPRETATIONS: The Administrative Board shall be the final authority on interpretations of this Agreement. However, this Agreement is not intended to prohibit or restrict any of the parties from seeking relief of disputes through the courts. This Agreement shall be construed by and governed by the laws of the State of Florida.

27. EFFECTIVE TERM: This Agreement shall take effect immediately upon its proper and complete execution by each Party and upon the filing of a copy of the same with the Clerk of the Circuit Court in and for Palm Beach County, and shall remain in full force and effect uless otherwise terminated pursuant to Paragraph 5 or 28.

28. TERMINATION: Except in the case of termination under Paragraph 5, this Agreement may be terminated upon forty-five (45) days written notice given by any party to the Administrative Board who shall within fifteen (15) days, serve written notice to all parties through their respective Clerk or Chief Administrative Officer.

INTERLOCAL EMERGENCY SERVICES AGREEMENT

IN WITNESS WHEREOF, the parties hereto have caused this Agreement to be entered into and executed the _____ day of _____ ,1994.

WITNESSES:

CITY, TOWN, VILLAGE OR
COUNTY OF _____

_____

_____

by

Mayor/Chairperson

_____

_____

by
CITY, TOWN, VILLAGE OR
COUNTY
Manager/Administrator

(CORPORATE SEAL)

ATTEST:

_____
Clerk

Approved as to form:

_____
City, Village, Town or County Attorney

STATE OF FLORIDA:
COUNTY OF PALM BEACH:

BEFORE ME, an officer duly authorized by law to administer oaths and take acknowledgments, personally appeared _____ Mayor, City Manager and City Clerk, respectively, of the City of _____ Florida, a municipal corporation of Florida, and acknowledged they executed the foregoing Agreement as the proper officials of the City of _____ , and the same is the act and deed of the City of

_____

THE FOREGOING, I have set my hand and official seal at _____, In the State and County aforesaid on the _____ day of _____ ,19 ____

(SEAL)

_____
Notary Public
My Commission Expires:

19

*Fire and*
*Rescue Division*
*FIRESCOPE*

California Fire Service and Rescue
Emergency Mutual Aid System

# Fire and Rescue
# Mutual Aid System

**Pete Wilson**
Governor

**Richard Andrews, Ph.D.**
Director
Governor's Office of Emergency Services

# MUTUAL AID PLAN

NEIL R. HONEYCUIT, Chief
Fire and Rescue Branch

Word Processing: Lena Webb

Printing/Assembly: Sue Dubie-Holbrook

Editing: Fire and Rescue Service Advisory
Committee/FIRESCOPE Board of Directors

# CALIFORNIA FIRE SERVICE AND RESCUE EMERGENCY MUTUAL AID PLAN

## TABLE OF CONTENTS

# CALIFORNIA FIRE SERVICE AND RESCUE
# EMERGENCY MUTUAL AID PLAN

## I. INTRODUCTION

The *California Fire Service and Rescue Emergency Mutual Aid Plan* is an extension of, and supportive document to, the California Emergency Plan. The California Emergency Plan is published in four parts as follows:

Part One:      BASIC EMERGENCY PLAN

Part Two:      PEACETIME EMERGENCY PLAN

Part Three:    COMPENDIUM OF LEGISLATION AND REFERENCES

Part Four:     WAR EMERGENCY PLAN

Parts One, Two and Four provide the planning basis and concepts for the development of the *California Fire Service and Rescue Emergency Mutual Aid Plan*. This Plan supports the concepts of the Incident Command System (ICS), the Integrated Emergency Management System (IEMS), and multi-hazard response planning. It is intended that more detailed operational plans will supplement this document at the local, area, and regional levels. California fire and rescue service conducts emergency operations planning at four levels: Local, Operational Area, Regional, and State. To effectively implement the plans formulated at the various levels, all plans should be developed within the framework of the *California Fire Service and Rescue Emergency Mutual Aid Plan*.

The *California Fire Service and Rescue Emergency Mutual Aid Plan* as we know it today, was first prepared and adopted in 1950 as Annex 3-C of the California State Civil Defense and Disaster Relief Plan. This plan has been reviewed, revised, approved, and adopted after careful consideration by the OES Fire and Rescue Service Advisory Committee/FIRESCOPE Board of Directors.

A.     **PURPOSE OF THE PLAN:**

1.     To provide for systematic mobilization, organization and operation of necessary fire and rescue resources of the state and its political subdivisions in mitigating the effects of disasters, whether natural or man-caused.

2.     To provide comprehensive and compatible plans for the expedient mobilization and response of available fire and rescue resources on a local, area, regional and statewide basis.

3.     To establish guidelines for recruiting and training auxiliary personnel to augment regularly organized fire and rescue personnel during disaster operations.

4.     To provide an annually-updated fire and rescue inventory of all personnel, apparatus and equipment in California.

5.  To provide a plan and communication facilities for the interchange and dissemination of fire and rescue-related data, directives, and information between fire and rescue officials of local, state, and federal agencies,

6.  To promote annual training and/or exercises between plan participants.

## B.  PLANNING BASIS:

1.  No community has resources sufficient to cope with any and all emergencies for which potential exists.

2.  Fire and rescue officials must preplan emergency operations to ensure efficient utilization of available resources.

3.  Basic to California's emergency planning is a statewide system of mutual aid in which each jurisdiction relies first upon its own resources.

4.  The California Disaster and Civil Defense Master Mutual Aid Agreement between the State of California, each of its counties, and those incorporated cities and fire protection districts signatory thereto:

    a.  Creates formal structure for provision of mutual aid;

    b.  Provides that no party shall be required to unreasonably deplete its own resources in furnishing mutual aid;

    c.  Provides that the responsible local official in whose jurisdiction an incident requiring mutual aid has occurred shall remain in charge at such incident, including the direction of such personnel and equipment provided through mutual aid plans pursuant to the agreement;

    d.  Provides the intra- and inter-area and intra-regional mutual aid operational plans shall be developed by the parties thereof and are operative as between the parties thereof in accordance with the provisions of such operational plans;

    e.  Provides that reimbursement for mutual aid extended under this agreement and the operational plans adopted pursuant thereto, shall only be pursuant to the state law and policies, and in accordance with Office of Emergency Services policies and procedures.

5.  The state is divided into six mutual aid regions to facilitate the coordination of mutual aid. Through this system the Governor's Office is informed of conditions in each geographic and organizational area of the state, and the occurrence or imminent threat of disaster.

6.     In addition to fire and rescue resources, emergency operations plans should include both public and private agencies with support capability and/or emergency operation responsibilities.

7.     Emergency operations plans should be distributed to, and discussed with, management, command, operational and support level personnel within each planning jurisdiction,

8.     Emergency operations plans must be continuously reviewed, revised, and tested to encompass change and refinement consistent with experience gained through disaster operations and training, and changes in resource availability.

9.     Emergency operations plans are to be reviewed, revised, and updated every five years.

This *California Fire Service and Rescue Emergency Mutual Aid Plan* supersedes the Fire and Rescue Emergency Plan, revised June 1978.

## II. AUTHORITIES

**A.**     California Emergency Services Act (Chapter 7 of Division 1 of Title 2 of the Government Code) 1970 Statutes.

**B.**     California Master Mutual Aid Agreement.

**C.**     Labor Code, State of California (Section 3211.92, Disaster Service Worker).

**D.**     Government Code, State of California (Section 8690.6).

## III. REFERENCES

**A.**     Federal Civil Defense Guide (Part E, Chapter 10, with Appendixes 1 and 2, Fire Prevention and Control during Civil Defense Emergencies).

**B.**     Governor's Executive Order No. D-25.

**C.**     Governor's Administrative Orders for State Agencies.

**D.**     Public Resources Code.

**E.**     Office of Emergency Services, Multihazard Functional Planning Guidance.

**F.**     National Interagency Incident Management System.

**G.**     Incident Command System.

**H.**     Multi-agency Coordination System.

## IV. DEFINITIONS

**A.**     **Fire and Rescue Resources:**

California fire and rescue resources shall include, but not be limited to, the necessary personnel, apparatus and equipment under the direct control of the fire and rescue service needed to provide mutual aid assistance for all emergencies; i.e., fire engines, ladder trucks, emergency medical service units, hazardous materials units, search and rescue, crash fire rescue, bulldozers, helicopters, fixed wing aircraft, hand crews, fire boats, communications equipment, etc.

B. **Local Emergency:**

Shall mean the existence of conditions within the territorial limits of a local agency, in the absence of a duly proclaimed state of emergency, which conditions are a result of an emergency created by great public calamity such as air pollution, extraordinary fire, flood, storm, earthquake, civil disturbances or other disaster which is or is likely to be beyond the control of the services, personnel, equipment and facilities of that agency and require the combined forces of other local agencies to combat. (California Emergency Services Act, Chapter 7 of Division 1 of Title 2 of the Government Code - 1970 Statutes.)

C. **State of Emergency:**

Means the duly proclaimed existence of conditions of extreme peril to the safety of persons and property within the state caused by such conditions as air pollution, fire, flood, storm, civil disturbances or earthquake, or other conditions, except as a result of war-caused emergencies, which conditions by reason of their magnitude, are or are likely to be beyond the control of 'the services, personnel, equipment and facilities of any single county, city and county, or city, and would require the combined forces of a mutual aid region or regions to combat. "State of Emergency" does not include, nor does any provision of this plan apply to any condition resulting from a labor controversy. (California Emergency Services Act, Chapter 7 of Division 1 of Title 2 of the Government Code - 1970 Statutes.)

D. **State of War Emergency:**

Means the conditions which exists immediately, with or without a proclamation thereof by the Governor, whenever this state or nation is attacked by an enemy or upon receipt by the state of a warning from the federal government indicating that such attack is probable or imminent. (California Emergency Services Act, Chapter 7 of Division 1 of Title 2 of the Government Code - 1970 Statutes.)

E. **Disaster Service Worker:**

Means any person who is registered with a disaster council for the purpose of engaging in disaster service pursuant to the "California Emergency Services Act" without pay or other consideration. "Disaster Service Worker" includes volunteer civil defense workers and public employees and also includes any unregistered person impressed into service during a State of Emergency by a person having authority to command the aid of citizens in the execution of that person's duties. "Disaster Service Worker" does not include any person registered as an active fire and rescue service member of any regularly-organized volunteer fire department, having official recognition and full or partial support of the county.

Pursuant to the *Califorinia Fire Service and Rescue Emergency Mutual Aid Plan,* "Disaster Service Workers" shall be recruited and trained to augment the regular fire and rescue forces. They will assist in fighting fires and/or rescuing persons, and save property and perform other duties as required.

Training necessary to engage in such activities is defined as authorized and supervised training carried on in such a manner and by a qualified person as the local disaster council shall prescribe. (Section 3211.92, California Labor Code.)

**F.    Mutual Aid:**

An agreement in which two or more parties agree to furnish resources and facilities and to render services to each and every other party of the agreement to prevent and combat any type of disaster or emergency.

Local needs not met by the *California Fire Service and Rescue Emergency Mutual Aid Plan* should be resolved through development of local <u>automatic,</u> or <u>mutual aid</u> agreements.

1.    <u>Voluntary Mutual Aid</u>

Mutual aid is voluntary when an agreement is initiated either verbally or in writing. When in writing, which is preferable, conditions may be enumerated as to what and how much of a department's resources may be committed.

2.    <u>Obligatory Mutual Aid</u>

Mutual aid under a "State of War Emergency" shall be deemed obligatory. Mutual aid under a "State of Emergency" may be obligatory. (Emergency Services Act, 1970.)

3.    <u>Master Mutual Aid Agreement</u>

An agreement made and entered into by and between the State of California, its various departments and agencies, and the various political subdivisions, municipal corporations, and other public agencies of the State of California to facilitate implementation of Chapter 7 of Division 1 of Title 2 of the Government Code entitled "California Emergency Services Act."

**G.    Mutual Aid Region:**

A subdivision of the state's fire and rescue organization, established to facilitate the coordination of mutual aid and other emergency operations within a geographical area of the state, consisting of two or more county operational areas.

**H.    Operational Area:**

An intermediate level of the state fire and rescue organization, normally consisting of a county and all fire and rescue organizations within the county.

I.      **Assistance by Hire:**

Assistance by hire resources are those elements of personnel and equipment which are provided by cooperating agencies through specific arrangements not associated with this plan. Where such arrangements exist, parties should be thoroughly familiar with, and aware of, provisions at time of request and response.

## V. MUTUAL AID REGIONS

State of California

OFFICE OF EMERGENCY SERVICES

# MUTUAL AID REGIONS

Revised  09/88

## VI. POLICIES

The following policies form the basis of the *California Fire Service and Rescue Emergency Mutual Aid Plan:*

A. The basic tenets of emergency planning are self-help and mutual aid.

B. Emergency planning and preparation is a task which must be shared by all political subdivisions and industries as well as every individual citizen.

C. The *California Fire Service and Rescue Emergency Mutual Aid Plan* provides a practical and flexible pattern for the orderly development and operation of mutual aid on a voluntary basis between cities, cities and counties, fire districts, special districts, county fire departments, and applicable state agencies. Normal fire department operating procedures are utilized, including day-to-day mutual aid agreements, and plans which have been developed by local fire and rescue officials.

D. Operational Area and Region Plans shall be consistent with policy of the Master Mutual Aid Agreement and the *California Fire Service and Rescue Emergency Mutual Aid Plan.*

E. Reimbursement for mutual aid may be provided pursuant to a governor's disaster proclamation or when conditions warrant invoking the USFS/CDF/OES Cooperative Agreement. There is no other existing provision for mutual aid reimbursement.

   1. The Office of Emergency Services shall be required to provide direction, ongoing guidance and monitoring throughout the process until reimbursement is received by local agencies.

   2. Memorandums of understanding between federal, state and local agencies will not include a commitment of local resources without the expressed, written consent of the local jurisdiction(s).

F. In developing emergency plans, provisions should be made for integrating fire and rescue resources into mutual aid organizations for both fire and non-fire related disasters; i.e., earthquake, flood, radiological defense, hazardous materials incidents, and war-related sheltering and/or relocation of significant portions of the population. In planning for war-related emergencies, provisions for pre- and post-attack activities should be included; i.e., shelter improvement, radiological monitoring and decontamination.

G. In developing local mutual aid and emergency preparedness plans, provisions must be made for liability and property damage insurance coverage on apparatus and equipment used beyond the territorial limits of the political subdivision. Consideration must also be given to the rights, privileges, and immunities of paid, volunteer, and auxiliary personnel in order that they may be fully protected while performing their duties under a mutual aid agreement or an emergency preparedness plan. Provision is made in state laws to deal with these matters, and the procedure outlined therein should be followed to ensure maximum protection.

H.   Local mutual aid and emergency preparedness plans should reference the Master Mutual Aid Agreement by signature of all parties concerned.

I.   The State of California provides Workers' Compensation coverage for certain classes of auxiliary and volunteer personnel engaged in activities directly related to defense preparedness or disaster operations. Coverage is also extended to those unregistered persons impressed into service during a State of Emergency or State of War Emergency by a person having authority to command the aid of citizens in the execution of required duties. No payment of premium is required of local political subdivisions for such coverage. Coverage is not, however, extended to any member registered as an active fire fighting member of any regularly organized volunteer fire department having official recognition, and full or partial support of the county, city, town, or district in which such fire agency is located.

J.   Responsible Agency will:

1.   Reasonably exhaust local resources before calling for outside assistance;

2.   Render the maximum practicable assistance to all emergency-stricken communities under provisions of the Master Mutual Aid Agreement;

3.   Provide a current annual inventory of all fire department personnel, apparatus and equipment to the Operational Area Fire and Rescue Coordinator;

4.   Provide for receiving and disseminating information, data, and directives;

5.   Conduct the necessary training to adequately perform their functions and responsibilities during emergencies.

## VII.   ASSUMPTIONS

### A.   MAJOR EMERGENCIES:

Fire and rescue emergencies may reach such magnitude as to require mutual aid resources from adjacent local and state levels.

### B.   NATURAL DISASTER:

Natural disasters may necessitate mobilization of fire and rescue resources for the preservation and protection of life and property from threats other than fire; i.e., earthquake, flood, windstorm, etc.

### C.   SABOTAGE:

Fire sabotage is an enemy capability. Urban areas would be particularly subject to sabotage during pre-attack periods. Metropolitan and wildland areas would be especially vulnerable to incendiarism.

D.   **CIVIL DISTURBANCE:**

Civil disturbances frequently result in injuries to persons and property damage. Explosives and fire bombs are not uncommon components of civil disturbances. Fire and rescue mutual aid resources are likely to be mobilized for such occurrences.

E.   **POLITICAL VIOLENCE AND TERRORISM:**

Incidents of kidnaping, bombing, bomb threat and incendiarism to achieve political concession and public notoriety are becoming more prevalent,. Such terrorist and violent activity may result in fire and/or rescue emergencies necessitating mobilization of mutual aid resources. There is potential for immobilization of local resources through bombing, blackmail or sniping activity.

F.   **ATTACK:**

An enemy attack upon California or adjacent states could result in fire and rescue problems of such magnitude as to require utilization of all fire and rescue resources within the state and the exchange of resources between states.

G.   **LOCAL FIRE SERVICES:**

Local officials will maintain fire and rescue resources consistent with anticipated needs. Such services will be augmented by training volunteers for utilization in major disaster operations.

## VIII. ORGANIZATION

The fire and rescue service includes all public and private entities furnishing fire protection within the state. During a State of War, or when ordered by the Governor pursuant to the California Emergency Services Act, all such fire protection agencies become an organizational part of the Office of Emergency Services, Fire and Rescue Division.

A.   **LOCAL FIRE OFFICIAL:**

The fire chief, or senior fire and rescue official by other designated title, of each local entity providing public fire protection, will serve as fire and rescue representative to their respective Operational Area Fire and Rescue Coordinator.

B.   **OPERATIONAL AREA FIRE AND RESCUE COORDINATOR:**

Operational Area Fire and Rescue Coordinators are selected by the fire chiefs of local fire and rescue entities within an operational area. They shall each appoint one or more alternate fire and rescue coordinators. They, or their alternates, will serve on the staff of the Operational Area emergency services official in their respective area.

C.    **REGIONAL FIRE AND RESCUE COORDINATOR:**

Regional Fire and Rescue Coordinators are selected for a three-year term by Operational Area Fire and Rescue Coordinators within their respective regions. They shall each appoint one or more alternate Regional Fire and Rescue Coordinators. They, or their alternates, will serve on the staff of the OES Regional Manager during a State of War Emergency or State of Emergency proclaimed by the Governor.

D.    **STATE FIRE AND RESCUE COORDINATOR:**

The State Fire and Rescue Coordinator is the Chief of the Fire and Rescue Division of the Office of Emergency Services and is a staff member of the Director of the Office of Emergency Services. The State Fire and Rescue Coordinator is responsible for taking appropriate action on requests for mutual aid received through Regional Fire and Rescue Coordinator channels.

E.    **OTHER STATE AGENCIES:**

The Governor may assign to state agencies any activities concerned with the mitigation of the effects of an emergency (Article 7, Chapter 7 of Division 1 of Title 2 of the Government Code).

F.    **OFFICE OF EMERGENCY SERVICES:**

Provides coordination, guidance and assistance in planning, response and recovery to all disasters within the state,

G.    **CALIFORNIA DEPARTMENT OF FORESTRY AND FIRE PROTECTION:**

Provides fire protection services, and when available, rescue, first aid and other emergency services to those forest and other wildland areas for which the state is responsible, and to those areas and/or communities for which the state is responsible by contractual arrangements. The Department of Forestry and Fire Protection assists with personnel and equipment, including conservation camp crews (provided by the California Department of Corrections and California Youth Authority), in fire suppression, rescue and cleanup, communications, radiological monitoring, and personnel care as emergencies may require and dependent upon their normally assigned fire protection responsibilities.

H.    **STATE FIRE MARSHAL:**

Assists OES Fire and Rescue Division by providing personnel to facilitate coordination of mutual aid fire and rescue operations; provides personnel for arson and explosion investigation, flammable liquid pipeline emergencies.

I.    **DEPARTMENT OF FISH AND GAME:**

Assists other agencies in search and rescue missions; provides recommendations and guidelines for hazardous substance incidents which have or may contaminate streams, waterways, or state properties.

J.   **MILITARY DEPARTMENT:**

At the direction of the Governor, assists civil authorities in protecting life and property from fires, and conducts support operations designed to minimize devastation by fire; i.e., communications, transportation, evacuation, and engineering assistance, and by providing personnel and equipment for rescue operations.

K.   **DEPARTMENT OF TRANSPORTATION (CAL TRANS):**

Assists in the identification and containment of hazardous materials incidents and the coordination of traffic flow restoration with the Highway Patrol.

L.   **CALIFORNIA HIGHWAY PATROL:**

Acts as scene manager in hazardous material emergencies on all freeways, state owned toll bridges, highways, and roads in unincorporated areas of the state.

M.   **CALIFORNIA CONSERVATION CORPS**

Provides handcrews, helitack crews and fire camp crews to the California Department of Forestry and Fire Protection.

# FIRE AND RESCUE SERVICE ORGANIZATION

## STATE OF EMERGENCY

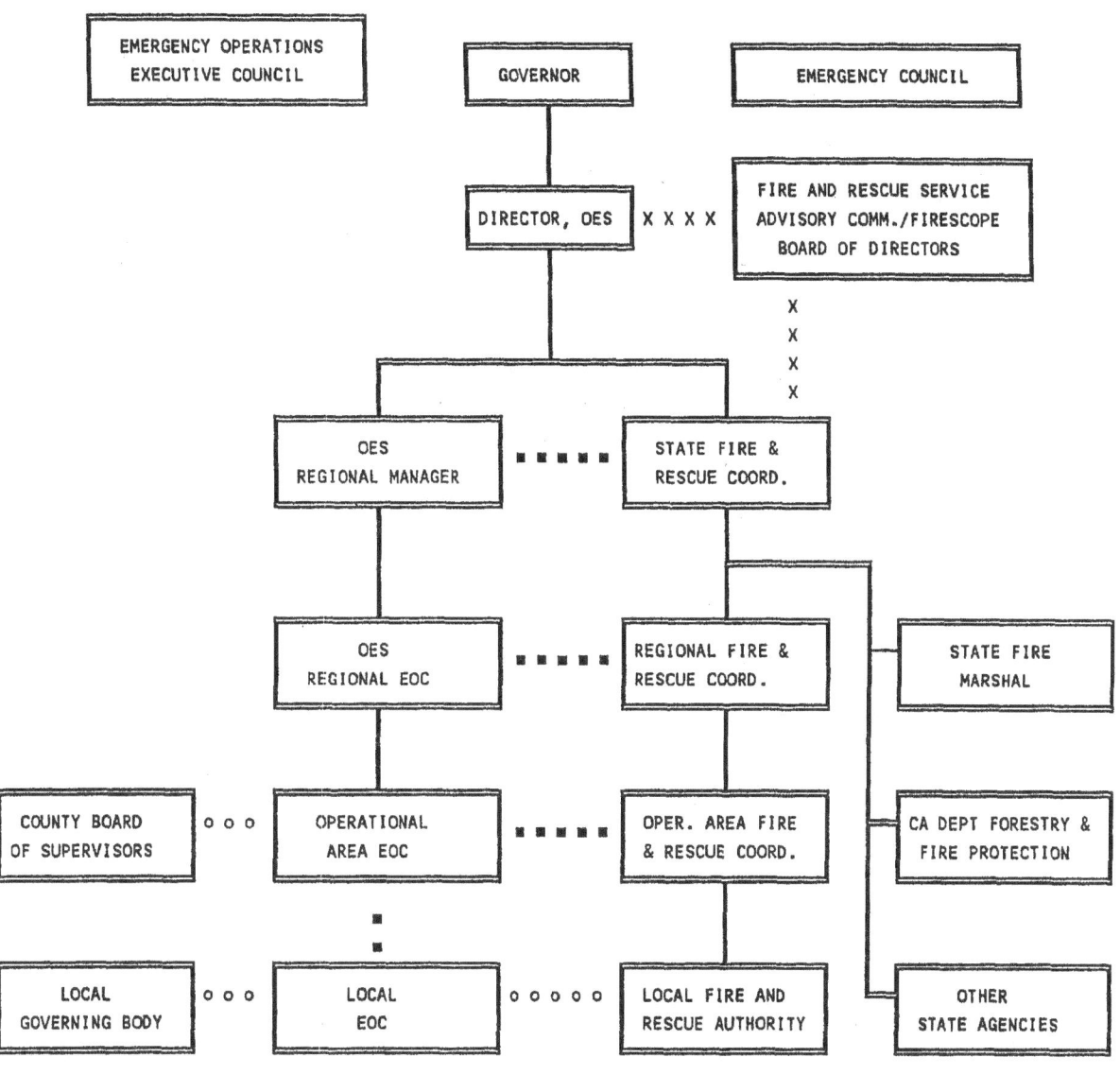

DIRECTION AND CONTROL
X X X X X X X ADVISORY
■ ■ ■ ■ ■ ■ ■ COORDINATION AND SUPPORT
o o o o o o o o LOCAL DIRECTION AND COORDINATION

# FIRE AND RESCUE SERVICE ORGANIZATION

## L O C A L   E M E R G E N C Y

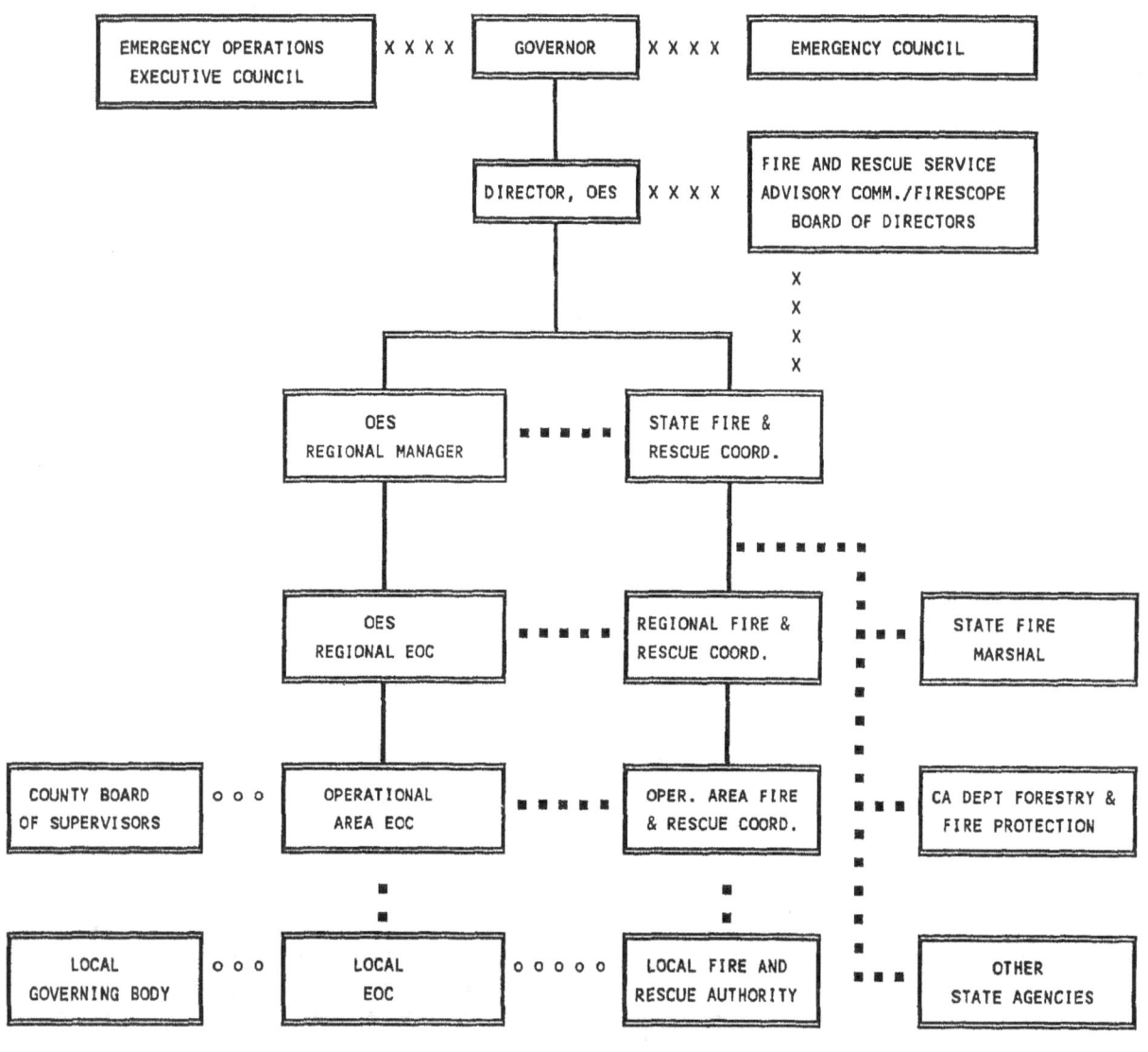

──────────── DIRECTION AND CONTROL
X X X X X X X  ADVISORY
▪ ▪ ▪ ▪ ▪ ▪ ▪  COORDINATION AND SUPPORT
o o o o o o o o  LOCAL DIRECTION AND COORDINATION

## IX.   RESPONSIBILITIES

### A.   LOCAL:

The appropriate Fire and Rescue Administrator:

1.   Directs all action toward stabilizing and mitigating the emergency, including controlling fires, saving lives, safeguarding property and assisting other emergency services in restoring normal conditions,

2.   Develops an effective emergency plan for use of the resources under its control and ensures that such a plan is integrated into the emergency plan of the operational area of which the fire and rescue administrator is a part. This plan should include provision for, but not be limited to, fire and rescue operations, earthquake, floods, civil disturbances, riots, bombings, industrial accidents, hazardous material incidents, mass casualty incidents, air and water pollution, etc.

3.   Makes maximum use of existing facilities and services within each community prior to requesting assistance from neighboring jurisdictions.

4.   Conducts mutual aid activities in accordance with established operational procedures.

5.   During emergency operations, keeps the Operational Area Fire and Rescue Coordinator informed on all matters.

6.   The agency receiving mutual aid is responsible for logistic support to all mutual aid personnel and equipment received.

7.   Prepares personnel and equipment inventories and forwards copies to the Operational Area Fire and Rescue Coordinator annually,

8.   Maintains an up-to-date schedule for alerting fire and rescue personnel in emergencies and a checklist of timely actions to be taken to put emergency operations plans into effect.

9.   Establishes emergency communications capabilities with the Operational Area Fire and Rescue Coordinator.

10.   Anticipates emergency needs for such items as emergency fire equipment, commonly used spare parts, and expendable supplies and accessories, and ensures functional availability of these in locations convenient for ready use.

11.   Develops a radiological monitoring capability and comprehensive training program within the department.

12.   When requesting aid, will be in charge of all manpower and apparatus received. Requests for mutual aid will be directed to the Operational Area Fire and Rescue Coordinator.

13. Provides mutual aid resources when requested by the Operational Area Fire and Rescue coordinator to the extent of their availability without unreasonably depleting their own resources.

14. Maintains appropriate records, data, and other pertinent information of mutual aid resources committed.

15. Provides approximate time commitment and justification of mutual aid needs in request for resources to the Operational Area Fire and Rescue Coordinator. Periodically evaluates the need of mutual aid committed and notifies the Area Coordinator.

**B.    OPERATIONAL AREA:**

The Operational Area Fire and Rescue Coordinator:

1. Organizes and acts as chairperson of an Operational Area Fire and Rescue Coordinating Committee composed of the Alternate Area Fire and Rescue Coordinators and/or fire chiefs within the operational area. The committee may include others as deemed necessary by the chairperson. This committee shall meet at least once each year and may hold additional meetings as deemed necessary by the chairperson.

2. In cooperation with its Operational Area Fire and Rescue Coordinating Committee, will:

    a. Organize, staff and equip area fire and rescue dispatch centers in accordance with the principles enumerated in the *California Fire Service and Rescue Emergency Mutual Aid Plan.*

    b. Select and submit to the Regional Fire and Rescue Coordinator the names of individuals to serve as the alternates at Operational Area fire and rescue dispatch centers.

    c. Aid and encourage the development of uniform fire and rescue operational plans within the Area.

    d. Aid and encourage the development of countywide fire and rescue communication nets operating on the approved fire frequency for the county. The communication net should tie the communications facilities of the county to the Operational Area Fire and Rescue Dispatch Center and alternate dispatch centers.

    e. Maintain an up-to-date inventory system on fire and rescue apparatus and personnel within the area for use in dispatching. Compile and forward this information to the respective Regional Fire and Rescue Coordinator annually.

    f. Develop a dispatching procedure for all state-owned OES fire apparatus, rescue trucks, and communication vehicles assigned within the area.

g.    Provide fire and rescue coordination to the OES operational area disaster preparedness official.

h.    Responsible to aid and assist local, region and state officials in planning, requesting, and utilizing mobilization centers as needed for staging strike teams and personnel where appropriate.

3.    During a "State of War Emergency," shall report to the area Emergency Operations Center to serve on the staff of the Operational Area Disaster Preparedness Director. An authorized representative may serve on this staff in place of the Operational Area Fire and Rescue Coordinator, if necessary.

4.    During a "State of Emergency" declared by the Governor, or as may be necessary, shall report to the area Emergency Operations Center or such other location as directed by the Regional Fire and Rescue Coordinator. If necessary, an authorized representative may assume this duty.

5.    Will be responsible for dispatching all OES and/or local fire and rescue resources within the operational area on major mutual aid operations.

a.    If the emergency is within the jurisdiction of the Operational Area Fire and Rescue Coordinator and overloads the communication facilities, it assigns dispatching of mutual aid equipment to an alternate fire and rescue dispatch center.

b.    Shall keep the Regional Fire and Rescue Coordinator informed of all operations.

c.    Evaluates requests for assistance from local agency; determines the resources from that operational area which can provide the most timely assistance, and initiates appropriate response thereof. Determines if the most timely assistance is from one adjacent operational area and if so, request assistance from that Operational Area Fire and Rescue Coordinator not to exceed five engine companies or individual resources, and notifies the Regional Fire and Rescue Coordinator of this action. When resources are needed from more than one adjacent area, either for timely response or when the need is beyond operational area capability, the request must be made to region.

d.    Determines approximate time commitment and justification of resources issued to local agency, and the length of time it will utilize these resources. Periodically evaluate the justification and commitment to the local agency of these resources, and notify the region.

e.    The Operational Area Fire and Rescue Coordinator will advise the requesting jurisdiction of the origin of resources responding to the request for assistance.

f.   Shall notify and advise the Regional Fire and Rescue Coordinator, in a timely manner, of the need to establish mobilization centers and/or staging areas.

6.   The Operational Area Fire and Rescue Coordinator is not responsible for any direct fire or other emergency operations except those which occur within the jurisdiction of its own department, agency, etc. The local official in whose jurisdiction the emergency exists shall remain in full charge of all fire and rescue resources, staffing, and equipment furnished for mutual aid operations.

## C.   REGIONS:

The Regional Fire and Rescue Coordinator:

1.   Organizes and acts as chairperson of a Regional Fire and Rescue Coordinating Committee, composed of Alternate Regional Fire and Rescue Coordinators and the Operational Area Fire and Rescue Coordinators within the region. This committee may include others as deemed necessary by the chairperson. This committee shall meet at least once each year and may hold additional meetings as deemed necessary by the chairperson.

2.   On receipt of information of an emergency within the region which may require regional mutual aid, or upon request of the State Fire and Rescue Coordinator, assumes its responsibilities for coordination and dispatch of regional mutual aid resources.

3.   In cooperation with its Fire and Rescue Coordinating Committee, will:

a.   Organize, staff, and equip a Regional Fire and Rescue dispatch center in accordance with the principles enumerated in the *California Fire Service and Rescue Emergency Mutual Aid Plan.*

b.   Select and submit to the State Fire and Rescue Coordinator, the names of individuals to serve as its alternates at the Regional Fire and Rescue dispatch centers.

c.   Aid, encourage, and approve the development of uniform fire and rescue emergency plans within the region, through the Operational Area Fire and Rescue Coordinators.

d.   Aid and encourage the development of countywide fire and rescue communication nets, tying local fire departments to an Operational Area Fire and Rescue dispatch center.

e.   Maintain an up-to-date inventory system of fire and rescue apparatus and personnel within the region for use in dispatching. Compile and forward this information to the State Fire and Rescue Coordinator annually.

4. During a "State of War Emergency," the Regional Fire and Rescue Coordinator or the authorized representatives shall report to the Regional Emergency Control Center, acting as Fire and Rescue liaison to the OES Regional Manager.

5. During a "State of Emergency" proclaimed by the Governor, or as may be necessary, the Regional Fire and Rescue Coordinator or the alternate will report to the Regional Emergency Control Center or other locations as directed by the State Fire and Rescue Coordinator.

6. Is responsible for dispatching all OES and/or local fire and rescue resources within. the region on major mutual aid operations.

   a. If the emergency exists within the jurisdiction of the Regional Fire and Rescue Coordinator and overloads the communication facilities, the Regional Fire and Rescue Coordinator assigns dispatching of mutual aid equipment to an Alternate Regional Fire and Rescue dispatch center.

   b. Keeps the State Fire and Rescue Coordinator informed of all operations within the region.

   C. Evaluates requests for assistance from Area; determines the region resources from that region which can provide the most timely assistance, and initiates appropriate response thereof. Determines if the most timely assistance is from an adjacent region and if so, requests assistance from that Region Fire and Rescue Coordinator (not to exceed five engines or individual resources), and must notify the State Fire and Rescue Coordinator of this action. When resources are needed from more than one adjacent region, either for timely response or when the need is beyond region capability, the request must be made to the State.

   d. Needs to request approximate time commitment and justification of resources issued to operational area, and length of time it will utilize these resources. Shall periodically evaluate the justification and commitment to the Operational Area of these resources, and notify the State.

   e. The Regional Fire and Rescue Coordinator will advise the requesting Area of the source of all assistance responding to the Area.

   f. Shall notify and advise the State Fire and Rescue Coordinator, in a timely manner, of the need to establish mobilization centers and/or staging areas.

7. Regional Fire and Rescue Coordinator will monitor and coordinate backup coverage within an area or region when there is a shortage of resources.

8. Calls and conducts elections within the respective Operational Areas for Operational Area Fire and Rescue Coordinator. These elections

will be held every three years and when a vacancy occurs or at the request of the State Fire and Rescue Coordinator. Communications and dispatch requirements will be considered in electing coordinators.

9. The Regional Fire and Rescue Coordinator is not responsible for any direct fire or other emergency operations except those which occur within the jurisdiction of its own department, agency, etc. The local official in whose jurisdiction the emergency exists shall remain in full charge of all fire and rescue resources furnished for mutual aid operations.

10. Responsible to aid and assist in planning, utilizing, and requesting mobilization centers as needed for staging strike teams during mutual aid operations.

## D. STATE:

The Chief, Fire and Rescue Division, Office of Emergency Services, is the State Fire and Rescue Coordinator.

### 1. Office of Emergency Services, Fire and Rescue Division:

a. Prepares, maintains, and distributes the basic *California Fire Service and Rescue Emergency Mutual Aid Plan* for coordinating statewide emergency fire and rescue resources which include, but are not limited to, all regularly established fire and rescue services within the state.

b. Develops and maintains a "Fire and Rescue Emergency Operations Plan" and "Standard Operating Procedure" for the use and dispatch of OES Fire and Rescue personnel, apparatus and other fire and rescue resources as necessary. Such plans shall be made available to appropriate levels of command; i.e., Operational Area and Region Fire and Rescue Coordinators, dispatch centers, and local fire and rescue officials.

c. Organizes, staffs and equips the State Fire and Rescue dispatch center and alternate facilities necessary to ensure effective statewide coordination and control of mutual aid fire and rescue operations.

d. Monitors ongoing emergency situations, anticipates needs, and prepares for use of inter-regional fire and rescue mutual aid resources, establishing priorities and authorizing dispatch.

e. State Fire and Rescue Coordinator will monitor and coordinate backup coverage between regions when there is a shortage of resources.

f. Consults with and keeps the Director of the Office of Emergency Services informed on all matters pertaining to the fire and rescue services, and through the State Fire and Rescue Coordinator, keeps the California Emergency Council informed of current policy matters and proposed revisions in the *California Fire Service and Rescue Emergency Mutual Aid Plan.*

g.  Consults with and assists federal and other state agency representatives on all matters of mutual interest to the fire and rescue service.

h.  Coordinates fire and rescue emergency mutual aid operations throughout the state, both on and off scene.

i.  Assists state and local fire and rescue agencies in utilizing federal assistance programs available to them and keeps them informed of new legislation affecting these programs.

j.  Assists in the coordination of the application and use of other state agency resources during a "State of Emergency" or "State of War Emergency."

k.  Develops and provides training programs and materials for effective application and utilization of the *California Fire Service and Rescue Emergency Mutual Aid Plan.*

l.  Encourages the development of training programs for specialized emergencies involving fire and rescue services; i.e., radiological monitoring, civil disturbances, staff and command training.

m.  Calls for and conducts elections for Regional Fire and Rescue Coordinator. These elections will be held every three years or any time a vacancy occurs.

n.  Develops procedures for reimbursement of state and local agency expenses associated with assistance rendered during a major incident.

p.  Standardizes forms and procedures for the records required for response of OES and/or local fire and rescue resources responding to incidents or operational area coverage which qualify for reimbursement.

2.  **California Department of Forestry and Fire Protection:**

a.  Provides fire protection services and when available, rescue, first aid and other emergency services to those forest and other wildland areas for which the state is responsible either directly or through contractual agreements, and to those areas and/or communities for which the Department of Forestry and Fire Protection is responsible by local government fire protection contracts.

b.  Provides supervision for adult conservation camp inmates, Youth Authority wards, and Conservation Center corpsmembers in fire defense improvement work, fire fighting, and other emergency activity.

c. Maintains a statewide radio and microwave communications system, extended throughout administrative districts and all counties in which the Department of Forestry and Fire Protection has a fire protection responsibility.

d. Has numerous agreements with federal, state and local jurisdictions providing for contract fire protection, assistance by hire and/or mutual aid.

e. The Department of Forestry and Fire Protection assists by:

(1) Maintaining and making available to the State Fire and Rescue Coordinator and Regional and Operational Area Fire and Rescue Coordinators, emergency operations plans and resource inventories of the Department's fire fighting equipment and personnel.

(2) Working cooperatively with State, Regional, and Operational Area Fire and Rescue Coordinators to integrate the Department's fire fighting resources into the state, regional, and local fire and rescue emergency mutual aid plan. (Any dispatch of CDF resources will be through CDF dispatch channels.)

(3) Providing personnel and equipment to the OES Fire and Rescue Division for rescue operations, including inmates, wards, and conservation corpsmembers under its jurisdiction. (Any dispatch of these crews under CDF jurisdiction will be through CDF dispatch channels.)

(4) Initiating requests to OES for federal fire suppression assistance under Section 417, Public Law 95-288, Disaster Relief Act of 1976, as the needs may arise because of wildland fires on state responsibility lands. Upon FDA approval, CDF will be responsible for working directly with the appropriate federal agency (US Forest Service) to secure assistance, keeping OES advised of this action and providing whatever information is required for justification and utilization of such assistance. (Pursuant to Governor's Administrative Order No. 75-20.)

3. **The State Fire Marshal:**

a. Assists OES Fire and Rescue Division by providing personnel to facilitate coordination of mutual aid fire and rescue operations, code enforcement, arson and explosion investigation, and flammable liquid pipeline emergencies.

b. Cooperates with OES Fire and Rescue Division in training Fire Marshal personnel for emergency operations.

C.   Assists OES and local jurisdictions in post-emergency damage surveys, building inspection, advising them on use and/or hazards of damaged facilities, including hazardous liquid pipelines and state-owned or occupied buildings. (Pursuant to Administrative Order No. 75-15.)

4.   **California Conservation Corps (75-5):**

Provides personnel and/or equipment: to assist in:

a.   The prevention and suppression of fire.

b.   Rescue of lost or injured persons.

C.   Support of other emergency operations.

5.   **California Highway Patrol (75-6):**

Provides assistance in:

a.   Emergency highway traffic regulations and control.

b.   Evacuation of residents/inhabitants.

C.   Scene manager for highway hazardous materials incidents. (V.C. 2454.)

6.   **Department of Corrections (75-10):**

a.   Supplies inmate personnel to support emergency operations,

b.   Provides congregate care for displaced persons at departmental facilities.

C.   Prepare food for consumption in the disaster area.

d.   Furnish emergency medical treatment to disaster victims.

7.   **Military Department (75-26):**

The Military Department may be activated by the Governor to provide any of the following support services:

a.   Air and surface transportation of authorized personnel, equipment, and supplies.

b.   Provision of interim voice, telegraph and teletype communications.

C.   Surface and aerial reconnaissance and photography.

d.   Mass feeding.

e.   Medical treatment and evacuation.

f.    Clearance of debris and rubble.

g.    Explosive ordinance disposal.

h.    Search and rescue.

i.    Emergency housing.

j.    Maintaining law and order.

The Military Department may respond directly to requests from the Department of Forestry and Fire Protection to aid in suppressing forest fires.

**8.    Department of Youth Authority (7540):**

a.    Ward camp crews assist in emergency operations.

b.    Provides congregate care for displaced persons at department facilities.

c.    Prepares food for consumption in the disaster area.

d.    Provides emergency medical treatment.

## X.    PROCEDURES - MUTUAL AID

Fire and rescue mutual aid rendered pursuant to California's Master Mutual Aid Agreement, is based upon an incremental and progressive system of mobilization. Mobilization plans have been based upon the concept of providing a local fire and rescue authority sufficient resources without extraordinary depletion of fire and rescue defenses outside the area of disaster. Under normal conditions, fire and rescue mutual aid plans are activated in ascending order; i.e., local, county, region, inter-region. Circumstances may prevail which make mobilization of significant fire and rescue forces from within the area or region of disaster impractical and imprudent. Inter-regional mutual aid is, therefore, not contingent upon mobilization of uncommitted resources within the region of disaster.

### A.    LOCAL FIRE AND RESCUE RESOURCES:

Local fire and rescue resources include resources available through automatic and/or day-to-day mutual aid agreements with neighboring jurisdictions. Local mobilization plans are activated by requests to participating agencies and must provide for notification of the

Operational Area (county) Fire and Rescue Coordinator upon activation. The Operational Area Fire and Rescue Coordinator must know of those resources committed under local plans when determining resource availability for subsequent response.

### B.    OPERATIONAL AREA FIRE AND RESCUE RESOURCES:

Operational Area Fire and Rescue resources are those which are made available to a participating agency through the approved and adopted

Operational Area (county) Fire and Rescue Emergency Mutual Aid Plan. Mobilization of Operational Area resources is activated by the Operational Area Fire and Rescue Coordinator, or his representative, in response to a request for assistance from an authorized fire and rescue official of the participating agency in need. The Operational Area Fire and Rescue Coordinator must notify the Regional Fire and Rescue Coordinator of area resources committed.

C.   **REGIONAL FIRE AND RESCUE RESOURCES:**

Regional fire and rescue resources include all resources available to a participating agency through the approved and adopted Regional Fire and Rescue Emergency Mutual Aid Plan. Operational Area (county) plans are significant elements of regional plans.

Mobilization of regional fire and rescue resources is activated by the Regional Fire and Rescue Coordinator in response to a request for assistance from an Operational Area Fire and Rescue Coordinator. Regional Fire and Rescue Coordinators must notify the Chief, OES Fire and Rescue Division, of resources committed.

D.   **INTER-REGIONAL FIRE AND RESCUE RESOURCES**

Inter-regional fire and rescue mutual aid is mobilized through the OES Fire and Rescue Coordinator in the afflicted mutual aid region. Selection of region(s) from which resources are to be drawn is made in consideration of the imminence of threat to life and property and conditions existing in the various regions. Fire and rescue forces will be mobilized in the strike team mode for inter-regional fire and rescue mutual aid response. Strike teams will normally consist of five engines and a qualified strike team leader unless unusual circumstances prevent assemblage in these numbers. (Each OES engine will be staffed by three ((3)) or more trained fire fighters.) Regional Fire and Rescue Coordinators must be notified of any strike team with less than five engines. This information must be relayed to the requesting agency. Strike teams of resources other than fire engines are identified within state ICS plans. Regional Fire and Rescue Coordinator requesting aid must specify the number, kind and type of strike teams and support resources desired. Utilization of Multi-Agency Coordination System resource ordering form (MACS Form 420) is required; procedures for use to be developed by OES.

Insofar as is practicable, an OES Assistant State Fire and Rescue Coordinator will be dispatched when five or more OES fire engines are activated. An OES fire equipment mechanic will be dispatched to fire emergencies in which OES fire equipment is involved when, in the opinion of the assigned OES Assistant State Fire and Rescue Coordinator, such response is needed.

OES fire and rescue resources may be included in local, area, and regional fire and rescue mutual aid plans when such resources are assigned within the boundaries covered by the plan.

E.   **DISPATCH CENTER:**

Fire and rescue dispatch centers must be carefully selected and adequately equipped for emergency operations. They should be located in a facility which conducts 24-hour-per-day operations. They must be equipped with facilities which permit direct communications with all fire and rescue agencies within their area of operation. They must be staffed with competent personnel and equipped with such maps, charts, records, and operational data necessary to perform emergency operations on a 24- hour-per-day, full-time basis, etc. Alternate fire and rescue dispatch centers should have the same capability as primary centers, thus ensuring continued operations in the event of failure of the primary centers.

F.  **TRAINING:**

1.  *The training of regular emergency personnel in specialized skills and techniques is essential if each level of the fire and rescue service is to successfully discharge assigned emergency responsibilities to handle all-risk emergencies.* Fire and rescue officials should identify key personnel with emergency assignments and ensure the adequacy of their training.

2.  The State of California Fire and Rescue Service has adopted the Incident Command System and Interagency Incident Management System. All agencies should maintain familiarity with these systems.

G.  **PLANNING:**

A well-developed, decision-making process can compliment all phases of mutual aid utilization. Failure to plan assures failure. Effective emergency action is dependent upon comprehensive planning. All mutual aid planning must consider the logistical and financial obligations incurred in either providing or receiving mutual aid assistance; i.e., fuel, feeding, overtime for personnel. Emergency situations evolve through a series of stages: Planning, Preparedness, Response and Recovery.

1.  Preparedness

    While this phase does not apply to all emergencies, involved jurisdictions, when possible, will put pre-emergency plans into operation. Such plans include alerting key personnel, ensuring readiness of essential resources, and preparing to move resources to the threatened area when required, If a request for mutual aid resources is anticipated, the next higher level of jurisdiction must be advised, including all available information relative to the expected threat, its location, imminence, potential severity, and other associated problems.

2.  Response

    The nature of emergency operations is dependent upon the characteristics and requirements of the situation. This phase may require the use of local, operational area, regional, and state resources. The magnitude and severity of fire and rescue service emergencies may develop rapidly and without warning. Equally rapid preplanned response on the part of the fire and rescue service is required. The situation may develop requiring federal assistance under provisions of a Presidential Disaster Declaration, thereby involving the Federal Emergency Management Agency (FEMA).

3.   Recovery

Planning for this phase should include reestablishment of essential public services, public safety inspections, and restoration of public facilities.

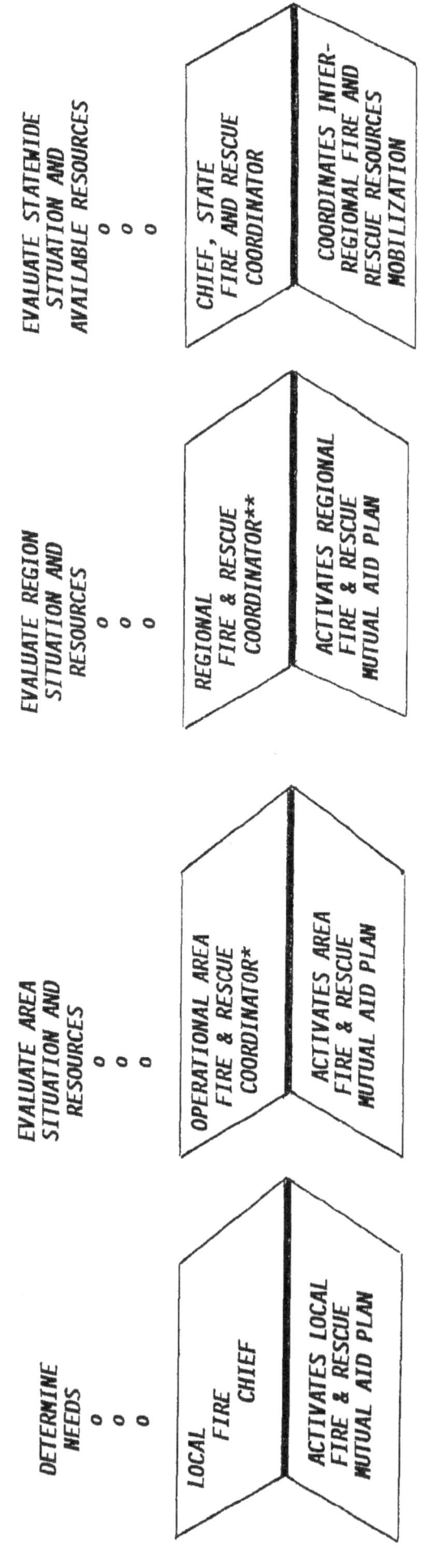

CHANNELS FOR REQUESTING FIRE AND RESCUE MUTUAL AID RESOURCES

DETERMINE NEEDS
o
o
o

LOCAL FIRE CHIEF

ACTIVATES LOCAL FIRE & RESCUE MUTUAL AID PLAN

EVALUATE AREA SITUATION AND RESOURCES
o
o
o

OPERATIONAL AREA FIRE & RESCUE COORDINATOR*

ACTIVATES AREA FIRE & RESCUE MUTUAL AID PLAN

EVALUATE REGION SITUATION AND RESOURCES
o
o
o

REGIONAL FIRE & RESCUE COORDINATOR**

ACTIVATES REGIONAL FIRE & RESCUE MUTUAL AID PLAN

EVALUATE STATEWIDE SITUATION AND AVAILABLE RESOURCES
o
o
o

CHIEF, STATE FIRE AND RESCUE COORDINATOR

COORDINATES INTER-REGIONAL FIRE AND RESCUE RESOURCES MOBILIZATION

\*    AREA BORDERLINE EMERGENCIES        (See Part IX., Responsibilities, Section B., Operational Area, 5.c. - Page 17)
\*\*  REGION BORDERLINE EMERGENCIES   (See Part IX., Responsibilities, Section C., Regions, 6.c. - Page 19)

Section 4 - Executive and Administrative Orders

# EXECUTIVE DEPARTMENT
## STATE OF CALIFORNIA

EXECUTIVE ORDER D-25-83

WHEREAS, it is the responsibility of the State of California to maintain a high degree of preparedness in the event of an earthquake, flood, fire, riot, epidemic, attack by a foreign power and other emergencies;

NOW, THEREFORE, I, GEORGE DEUKMEJIAN, Governor of the State of California, by virtue of the powers and authority vested in me by the Constitution and statutes of the State of California and in accordance with the provisions of Sections 8567 and 8595 of the Government Code, do hereby rescind Executive Order No. B-48-78, and do hereby issue this order to become effective immediately:

1. The Director, Office of Emergency Services, who is also the State Director of Emergency Planning and the State Director of Civil Defense, is responsible for preparation of the State of California Emergency Plan and the submission thereof, through the California Emergency Council, to me for approval;

2. The Director, Office of Emergency Services, shall coordinate the activities of all state agencies relating to preparation and implementation of the California Emergency Plan; and each state agency and officer shall cooperate with the Director and render all possible assistance during response and recovery phase of proclaimed emergencies;

3. The head of each department, bureau, board, commission and independent institution of state government, hereinafter referred to as an agency, is responsible for the emergency planning and preparedness of his or her agency;

4. Specific assignment of emergency functions to a given agency will be made in an Administrative Order by the Director, Office of Emergency Services, following consultation with the respective agency head;

5. Draft copies of agency emergency plans and procedures designed to carry out emergency assignments shall be submitted to the Director, Office of Emergency Services, for review and approval prior to publication;

6. Each agency shall prepare for and respond to emergency situations by ensuring:

   a. Protection of its personnel, equipment, supplies, facilities, and vital public records against the destructive forces of nature or man;

   b. The use of minimum resources required for continuation of normal services and redirection of all other resources to accomplish objectives in accordance with the California Emergency Plan;

C.   Designation of facilities for emergency use;

7.   Authority is hereby granted each state agency to properly train employees for emergency assignments in test exercises conducted by the agency or the Office of Emergency Services. Agency directors may allow compensation or compensating time off for training outside of regular working hours.

IN WITNESS WHEREOF I have hereunto set my hand and caused the Great Seal of the State of California to be affixed this 27th day of October

_George Deukmejian_
Governor of California

ATTEST:

_March Fong Eu_
Secretary of State

by _Anthony Miller_
Deputy Secretary of State

Part Three
Section 4
Page 2

5/1984

Revised 09/88
Page 30

## XII. MASTER MUTUAL AID AGREEMENT

There are references in the following agreement to the California Disaster Act, State Disaster Council, and various sections of the Military and Veterans Code.

Effective November 23, 1970, by enactment of Chapter 1454, Statutes 1970, the California Disaster Act (Sections 1500 ff., Military and Veterans Code) was superseded by the California Emergency Services Act (Sections 1550 ff., Government Code), and the State Disaster Council was superseded by the California Emergency Council.

Section 8668 of the California Emergency Services Act provides:

> "Master Mutual Aid Agreement" means the California Disaster and Civil Defense Master Mutual Aid Agreement, made and entered into by and between the State of California, its various departments and agencies, and the various political subdivisions of the state, to facilitate implementation of the purposes of this chapter.

Substantially, the same provisions as previously contained in Sections 1541, 1564, 1586 and 1587 of the Military and Veterans code, referred to in the foregoing agreement, are now contained in Sections 8633, 8618, 8652 and 8653, respectively, of the Government Code.

## CALIFORNIA DISASTER AND CIVIL DEFENSE
## MASTER MUTUAL AID AGREEMENT

This agreement made and entered into by and between the STATE OF CALIFORNIA, its various departments and agencies, and the various political subdivisions, municipal corporations, and other public agencies of the State of California;

## WITNESSETH:

WHEREAS, it is necessary that all of the resources and facilities of the State, its various departments and agencies, and all its political subdivisions, municipal corporations, and other public agencies be made available to prevent and combat the effect of disasters which may result from such calamities as flood, fire, earthquake, pestilence, war, sabotage, and riot: and

WHEREAS, it is desirable that each of the parties hereto should voluntarily aid and assist each other in the event that a disaster should occur, by the interchange of services and facilities, including, but not limited to, fire, police, medical and health, communication, and transportation services and facilities, to cope with the problems of rescue, relief, evacuation, rehabilitation, and reconstruction which would arise in the event of a disaster; and

WHEREAS, it is necessary and desirable that a cooperative agreement be executed for the interchange of such mutual aid on a local, countywide, regional, statewide, and interstate basis;

1.    Each party shall develop a plan providing for the effective mobilization of all its resources and facilities, both public and private, to cope with any type of disaster,

2.    Each party agrees to furnish resources and facilities and to render services to each and every other party to this agreement to prevent and combat any type of disaster in accordance with duly adopted mutual aid operational plans, whether heretofore or hereafter adopted, detailing the method and manner by which such resources, facilities, and services are to be made available and furnished, which operational plans may include provisions for training and testing to make such mutual aid effective; provided, however, that no party shall be required to deplete unreasonably its own resources, facilities, and services in furnishing such mutual aid.

3.    It is expressly understood that this agreement and the operational plans adopted pursuant thereto shall not supplant existing agreements between some of the parties hereto providing for the exchange or furnishing of certain types of facilities and services on a reimbursable, exchange, or other basis, but that the mutual aid extended under this agreement and the operational plans adopted pursuant thereto, shall be without reimbursement unless otherwise expressly provided for by the parties to this agreement or as provided in Sections 1541, 1586, and 1587, Military and Veterans Code; and that such mutual aid is intended to be available in the event of a disaster of such magnitude that it is, or is likely to be, beyond the control of a single party and requires the combined forces of several or all of the parties to this agreement to combat.

4.    It is expressly understood that the mutual aid extended under this agreement and the operational plans adopted pursuant thereto shall be available and furnished in all cases of local peril or emergency and in all cases in which a *STATE OF EXTREME EMERGENCY* has been proclaimed.

5.    It is expressly understood that any mutual aid extended under this agreement and the operational plans adopted pursuant thereto, is furnished in accordance with the "California Disaster Act" and other applicable provisions of law, and except as otherwise provided by law that: "The responsible local official in whose jurisdiction an incident requiring mutual aid has occurred shall remain in charge at such incident including the direction of such personnel and equipment provided him through the operation of such mutual aid plans." (Sec.1564, Military and Veterans Code.)

6.    It is expressly understood that when and as the State of California enters into mutual aid agreements with other states and the Federal Government, the parties to this agreement shall abide by such mutual aid agreements in accordance with the law.

7.    Upon approval or execution of this agreement by the parties hereto all mutual aid operational plans heretofore approved by the State Disaster Council, or its predecessors, and in effect as to some of the parties hereto, shall remain in full force and effect as to them until the same may be amended, revised, or modified. Additional mutual aid operational plans and amendments, revisions, or modifications of existing or hereafter adopted mutual aid operational plans, shall be adopted as follows:

a.    Countywide and local mutual aid operational plans shall be developed by the parties thereto and are operative as between the parties thereto in accordance with the provisions of such operational plans. Such operational plans shall be submitted to the State Disaster Council for approval. The State Disaster Council shall notify each party to such operational plans of its approval, and shall also send copies of such operational plans and who are in the same area and affected by such operational plans. Such operational plans shall be operative as to such other parties 20 days after receipt thereof unless within that time the party by resolution or notice given to the State Disaster Council, in the same manner as notice of termination of participation in this agreement, declines to participate in the particular operational plan.

b.    Statewide and regional mutual aid operational plans shall be approved by the State Disaster Council and copies thereof shall forthwith be sent to each and every party affected by such operational plans. Such operational plans shall be operative as to the parties affected thereby 20 days after receipt thereof unless within that time the party by resolution or notice given to the State Disaster Council, in the same manner as notice of termination of participation in this agreement, declines to participate in the particular operational plan.

c.    The declination of one or more of the parties to participate in a particular operational plan or any amendment, revision or modification thereof, shall not affect the operation of this agreement and the other operational plans adopted pursuant thereto.

d.    Any party may at any time by resolution or notice given to the State Disaster Council, in the same manner as notice of termination of participation in this agreement, decline to participate in any particular operational plan, which declination shall become effective 20 days after filing with the State Disaster Council.

e.    The State Disaster Council shall send copies of all operational plans to those state departments and agencies designated by the Governor. The Governor may, upon behalf of any department or agency, give notice that such department or agency declines to participate in a particular operational plan.

f.    The State Disaster Council, in sending copies of operational plans and other notices and information to the parties to this agreement, shall send copies to the Governor and any department or agency head designated by him; the chairman of the board of supervisors, the clerk of the board of supervisors, the County Disaster Council, and any other officer designated by a county; the mayor, the clerk of the city council, the City Disaster Council, and any other officer designated by a city; the executive head, the clerk of the governing body, or other officer of other political subdivisions and public agencies as designated by such parties.

8.    This agreement shall become effective as to each party when approved or executed by the party, and shall remain operative and effective as between each and every party that has heretofore or hereafter approved or executed this agreement, until participation in this agreement is terminated by the party. The termination by one or more of the parties of its participation in this agreement shall not affect the operation of this agreement as between the other parties thereto. Upon approval or execution of this agreement the State Disaster Council shall send copies of all approved and existing mutual aid operational plans affecting such party which shall become operative as to such party 20 days after receipt thereof unless within that time the party by resolution or notice given to the State Disaster Council, in the same manner as notice of termination of participation in this agreement, declines to participate in any particular operational plan. The State Disaster Council shall keep every party currently advised of who the other parties to this agreement are and whether any of them has declined to participate in any particular operational plan.

9.    Approval or execution of this agreement shall be as follows:

a.    The Governor shall execute a copy of this agreement on behalf of the State of California and the various departments and agencies thereof. Upon execution by the Governor a signed copy shall forthwith be filed with the State Disaster Council.

b.    Counties, cities, and other political subdivisions and public agencies having a legislative or governing body shall by resolution approve and agree to abide by this agreement, which may be designated as *"CALIFORNIA DISASTER AND CIVIL DEFENSE MASTER MUTUAL AID AGREEMENT."* Upon adoption of such a resolution, a certified copy thereof shall forthwith be filed with the State Disaster Council.

c.    The executive head of those political subdivisions and public agencies having no legislative or governing body shall execute a copy of this agreement and forthwith file a signed copy with the State Disaster Council.

10.    Termination of participation in this agreement may be effected by any party as follows:

a.    The Governor on behalf of the State and its various departments and agencies, and the executive head of those political subdivisions and public agencies having no legislative or governing body, shall file a written notice of termination of participation in this agreement with the State Disaster Council and this agreement is terminated as to such party 20 days after the filing of such notice.

b.    Counties, cities, and other political subdivisions and public agencies having a legislative or governing body shall by resolution give notice of termination of participation in this agreement and file a certified copy of such resolution with the State Disaster Council, and this agreement is terminated as to such party 20 days after the filing of such resolution.

*IN WITNESS WHEREOF* this agreement has been executed and approved and is effective and operative as to each of the parties as herein provided.

/s/ EARL WARREN
GOVERNOR

On behalf of the State of California and all its Departments and Agencies.

(SEAL)　ATTEST:

November 15, 1950

/s/ FRANK M JORDAN
SECRETARY OF STATE

# APPENDIX B: SAMPLE TECHNICAL RESCUE STANDARD OPERATING PROCEDURES

**Fairfax County (VA) Fire and Rescue Department**

**Metro-Dade County (FL) Fire Department**

FAIRFAX COUNTY FIRE & RESCUE DEPARTMENT

FAIRFAX, VIRGINIA

DEPARTMENT OPERATING MANUAL

TECHNICAL RESCUE

OPERATIONS DIVISION

JULY 1, 1992

2nd Edition

| FAIRFAX COUNTY FIRE AND RESCUE DEPARTMENT OPERATING MANUAL | |
| --- | --- |
| TITLE : TECHNICAL RESCUE MANUAL - 2nd EDITION | |
| DIVISION : OPERATIONS | |
| FIRE CHIEF : *Glenn A. Gaines* | DATE : JULY 1, 1992 |

## TABLE OF CONTENTS

# TABLE OF CONTENTS - continued

# I.   INRODUCTION

## A.   Purpose

The purpose of this manual is to establish standards for the organization, training, equipment and operations *of* the Technical Rescue Operations Team (TROT)

## B.   Authority

This Department Operating Manual (DOM) is issued under the signature of the Fire Chief, through the Operations Division, in accordance with Standard Operating Procedure (SOP) 1.0.01, Written Department Communications.

## C.   Objective

The objective *of* this manual is to provide personnel with a comprehensive reference document which establishes standards and operationally effective procedures, based on safety considerations and individual and functional responsibilities.

## D.   Safety

Technical rescue operations require expert skills and involve varying degrees of risk to personnel and the public depending on the situation.

Department personnel are *subject to performing* tasks during rescue operations ranging *from* tending a guide line to removing a sick or injured worker *from* a construction crane. In either case, a life may be at stake and them is no room for error. Personnel must be proficient in their performance *of* the practices and evolutions prescribed in this document.

## II.   ORGANIZATION

### A.  Team Composition and Responsibility

1.  Personnel

   a. The TROT shall consist of the following personnel:

   1) Personnel assigned to Engine 18, Rescue Squad 18, and Fire Stations 14 and 21.
   2) Personnel at the rank of fire technician - technical rescue.
   3) Up to six fire officers, two per shift, not assigned to a technical rescue station, as designated by the shift assistant chief.
   4) Personnel, of any rank and assignment, certified in Technical Rescue Operations - II (TRO-II).

   b. Team personnel are authorized to wear the team patch and/or insignia in accordance with SOP 2.9.08, Uniforms.

2.  Apparatus and Staffing

   a. The team shall staff and operate the following apparatus in accordance with SOP 1.1.03, Staffing Procedures and the provisions of this manual.

   1) Technical Rescue Squads 14, 18 and 21: One (TRO-II) fire officer (career), two technicians or firefighters - technical rescue. Engine 18 may respond with Squad 18.

   2) Technical Rescue Units 14 and 21(CI14 & CI21): Two, three 0: five personnel. The station officer-in-charge shall place the engine company and/or ambulance out-of-service to staff the Technical Rescue unit.

3.  Responsibility

   a. The team shall have the primary responsibility for operations on the following types of incidents:

   1) Confined Space Rescue
   2) Rope Rescue (Rope Operations - III)
   3) Structural Collapse Rescue
   4) Trench Rescue

b. The team may be dispatched to assist on the following types of incidents:

   1) Agricultural and Industrial Rescue
   2) Construction Site Rescue
   3) Transportation Rescue of a Unique Nature

## B. Alarm Assignment

1. Technical Rescue Task Force

   The TROT shall respond and operate as a Task Force. No portion of the Technical Rescue Task Force shall be placed in service on incidents which require the services or equipment of the team.

2. Technical Rescue Box

   a. The Public Safety Communications Center (PSCC) shall dispatch technical rescue apparatus in accordance with SOP 4.1.02, Response Plan, at the request of any unit officer, incident commander or at the discretion of the PSCC Uniformed Fire Officer (UFO).

   b. The Computer Aided Dispatch (CAD) System recommends a Technical Rescue Box on the following Event Type Codes. The complete apparatus complement varies according to the individual Event Type Code.

      1) CAVEIN: Trench cave-in, building collapse, mine shaft entrapment, or a vehicle accident involving a building, with entrapment
      2) METROF: Metrorail emergencies (collision or derailment)
      3) PTRANF: Train collision/derailment
      4) RABOVE: Aerial rescue
      5) RBELOW: Below-grade rescue

## C. Administrative Chain of Command

1. Deputy Chief - Operations Division (DFCO)

2. Battalion Chief - Technical Services Officer (TSO)

   a. The Technical Services Officer, under the direction of the Deputy Chief of Operations, is responsible for the organization, training, equipment and operations of the TROT and their participation in the United States Disaster Assistance Response Team (DART).

b.    The Technical Services Officer shall manage these responsibilities through the Technical Rescue Committee. The Technical Services Officer shall appoint the members of the committee, conduct meetings at least quarterly and provide the Deputy Chief of Operations with a report.

3.    Technical Rescue Committee

a.    Captain – Senior Technical Officer (1: 14, 18 or 21)

1)    The senior technical officer shall be responsible for the administrative and tactical operations of the TROT and shall serve as the Chairperson of the Technical Rescue Committee.

2)    The senior technical officer shall appoint one of the training officers as lead instructor for each calendar year.

b.    Staff Services Officer

1)    The staff services officer shall be responsible for all correspondence and the maintenance of a comprehensive computer based system of administrative records and documents.

2)    The staff services officer, with concurrence of the TSO, shall prepare a general order each November, listing the projected calendar year training schedule.

3)    The staff services officer shall notify personnel with expired TRO-II certification, by memorandum, each January.

4)    The staff services officer shall provide the technical services officer and the Technical Rescue Committee with a copy of the annual training record each January.

c.    Training Officers (a total of 3; one per shift)

1)    The training officers shall be responsible for planning, scheduling, conducting and monitoring training on their respective shifts. The lead training officer may designate the lead instructor for a specific course. The senior technical officer shall be responsible for reviewing

all drill outlines prior to training
taking place. This review is for quality
control. Each  shift training officer, or
designee, shall then conduct the training
on their  respective shift.

2) The training officers/lead instructors are
responsible for coordinating training
schedules and apparatus status
requirements with  their  shift's  lead
battalion chief.

3) The training officers/lead instructors are
responsible for ensuring that class
rosters are delivered to the Fire and
Rescue Academy and the Fire and Rescue
Academy Liaison Officer.

4) The training Officers shall also keep
members informed of their current status
and training needs to maintain
certifcation on a quarterly basis
beginning January 1.

d. Fire and Rescue Academy Liaison Officer

The Fire and Rescue Academy liaison officer
shall be responsible for coordinating training
issues and schedules with the training
officers. The Fire and Academy liaison
Officer is responsible for ensuring that class
rosters are delivered to the training officers
and the staff services officer..

e. Office of Foreign Disaster Assistance
(OFDA)/Federal Emergency Management Agency
(FEMA) Liaison Officer

1) The OFDA/FEXA liaison officer shall be
responsible for coordinating DART issues
and operations with the United States OFDA
and FEMA.

2) The OFDA/FEMA liaison officer shall manage
this responsibility through sub-
committees. The OFDA/FEMA liaison officer
shall appoint the members of the sub-
committees and  provide the Technical
Rescue Committee with progress reports.

f. Volunteer Liaison Officer

1) The Volunteer liaison officer shall be
responsible for coordinating volunteer
participation in the technical rescue
program

g. Special Projects Officer

   1) The special projects officer shall be responsible for the completion of various administrative projects assigned by the senior technical officer.

      a) The special projects officer shall publish the TROT Activity Report on a quarterly basis.

      b) The special projects officer shall publish and distribute the minutes of each technical rescue committee meeting.

      c) The special projects officer shall research changes in laws, regulations and standards that relate to team operations and report findings to the senior technical officer.

h. Equipment/Logistics Officer (2)

   1) The equipment/logistics officers are responsible for assisting with the management of the tools and equipment used by the TROT.

   2) The equipment/logistics officers shall maintain an inventory tracking system of all tools and equipment assigned to the OFDA/FEMA Response Team.

   3) The equipment/logistics officers shall identify maintenance requirements for existing tools and equipment and coordinate those needs with the senior technical officer.

   4) The equipment/logistics officers shall research information on needed tools and equipment to assist with the completion of TROT capital equipment budget requests.

# III. TRAINING

## A. Safety Regulation

1. Personnel shall abide by all safety regulations established in this manual, SOP 2.3.04, Personnel Safety, SOP 3.1.02, Protective Clothing, the Rope Operations Manual and the Incident Command System Manual on all incidents and training evolutions.

2. Technical Rescue instructors shall be certified in accordance with SOP 7.10.01, Adjunct Faculty, Instructor and Academy Duty Officer Qualifications.

## B. Courses

1. Technical Rescue Operations I (TRO-I)

   a. Personnel

      Operations Division personnel shall be certified in Technical Rescue Operations I.

   b. Contents

      TRO-I is a four-hour lecture and demonstration course which will provide personnel with an overview of the TROT's organization, training, equipment and operations.

   c. Delivery

      1) Recruit School
      2) Company Drills (Review)
      3) Operations Academy Rotations (OARS) (Review)

   d. Certification

      Fire and Rescue Academy Graduation Certificate (Recruit School).

2. Technical Rescue Operations II (TRO-II)

   a. Personnel

      1) The following personnel shall be certified in Technical Rescue Operations II:

         a)     Personnel assigned to Engine 18, Rescue Squad 18, and Fire Stations 14 and 21

b) Personnel at the rank of fire technician - technical rescue

c) Up to six fire officers, two per shift, not assigned to a technical rescue station, as dosignated by the shift assistant chief (relief officers)

d) Personnel assigned to the (OFDA/FEMA) United States Disaster Assistance Response Team (DART). This requirement does not apply to personnel functioning as staff support to the DART.

b. Contents

1) TRO-II is an 80-hour lecture and practical performance course which will provide personnel with the basic skills required to perform technical rescue operations safely and efficiently.

a) Confined Space Rescue ........ . 16 hrs
b) Rope Operations III .......... . 24 hrs
c) Structural collapse Rescue ... . 16 hrs
d) Trench Rescue............... . 24 hrs

c. Delivery

Technical Rescue Operations School (Fall-Odd Years)

d. Certification

1) Fire and Rescue Academy Certificate

a) Technical Rescue operations II

2) Certification Maintenance

a) Maintenance of TRO-II certification requires personnel to attend a minimum Of three Technical Rescue Refresher classes per calendar year.

b) Personnel who complete the TRO-II course shall receive credit for two refresher classes for the calendar year in which they attend.

3) Expired Certification

    a) Expired certification will prohibit personnel from being utilized as minimum staffing on a technical rescue squad. (Personnel holding the rank of Fire Technician-Technical Rescue who fall to maintain their certification will be subject to demotion.)

    b) Personnel desiring to renew their certification shall contact their training officer. Staff and volunteer personnel shall contact the lead training officer. The training officer will review the training records and may require the applicant to take written and/or practical exams prior to reinstatement.

3. Technical Rescue Refresher Classes

  a. Personnel

    1) Personnel assigned to Engine 18, Rescue Squad 18, Fire Stations 14 and 21, all personnel at the rank of fire technician - technical rescue and all designated relief officers shall attend refresher Classes while on duty.

    2) TRO-II personnel, Other than specified above, shall make arrangements with an alternate shift's training officer to attend refresher classes while off duty, without compensation.

  b. Contents

    1) Technical Rescue Refresher Classes are eight-hour courses which will provide personnel with a review of TRO-II standards and evolutions and/or present new material as changes in technology and techniques are developed.

    2) Refresher classes may periodically run 16 or 24 hours depending on the material being presented.

c.  Delivery

Technical Rescue Refresher Classes are
conducted    6 tlmes/year/shlft for an annual
total of  18 sessions. They are conducted
during alternate months, commencing in January
of each year. They are normally held at the
Fire and Rescue  Academy  and begin at  0830
hours. A calendar year training schedule is
published as a general order  each   November.

d.  Certification

TRO-II personnel are required to attend three
refresher classes/year (Chapter III.2.d.(2).

4.  Technical Rescue Standard Evolution Classes

a.  Personnel

1)  Personnel assigned to Engine 18, Rescue
    Squad 18, and Fire Stations 14 and 21
    shall participate in the Standard
    Evolution Classes while on duty.

2)  TRO-II personnel, other than specified
    above, are encouraged to rake arrangements
    with an alternate shift's training officer
    to attend Standard Evolution Classes while
    off duty, without compensation.

b.  Contents

Technical Rescue Standard Evolution Classes
are two hour courses which will provide
assigned personnel with a review of  TRO-II
standard evolutions.

c.  Delivery

Technical Rescue standard   Evolution Classes
are conducted 6 time/year/shift for an annual
total Of 18 session. They are conducted
during alternate months, commencing in
February of each year. They are normally hold
in each technical rescue station (14, 18  and
21) On the same-day. A calendar year training
schedule  is published as a general order each
November.

d.   Certification

Technical Rescue Standard Evolution Classes
are primarily designed to develop teamwork
among personnel assigned to technical rescue
apparatus and have no bearing on TRO-II
certification status.

5.   DART Training Sessions

a.   Personnel

1)   Personnel assigned to the OFDA/FEMA DART
shall attend DART training sessions as
prescribed by general order.

2)   The DART roster is limited to 50 TRO-II
personnel. The Fire Chief may appoint
additional staff personnel.

3)   Personnel desiring to attain membership in
the DART shall submit a letter of interest
to the OFDA/FEMA liaison officer, through
their battalion chief. Personnel are
appointed to the DART by the Technical
Services Officer in conjunction with the
Technical Rescue Committee.

b.   Contents

1)   DART training sessions are courses of
variable length designed  to provide team
numbers with information regarding the
organization, training, equipment and
operations of the team.

2)   Drills will be scheduled periodically by
OFDA and PEXA.

c.   Delivery

DART training session dates, times,
location(s), personnel, purpose and contents
will be issued as a general order or
informational bulletin.

d.   Certification

TRO-II personnel are required to maintain
their TRO-II Certification, Passport and
Inoculation Record, and attend DART training
sessions to remain on the team. Staff
personnel are required to maintain their
passport, inoculation records and attend DART
training sessions to remain on the team.

C.  Zones

    1.  The following zones have been established for the
       delivery of special training sessions as
       periodically required by the Deputy Chief of
       Operations.

       a.  Rescue Squad 14
          Stations:   5, 9, 11, 16, 19, 20, 22, 24, 27,
                    32 and 35

       b.  Rescue Squad 18
          Stations:   1, 8, 10, 13, 23, 26, 28, 29 and
                    30

       c.  Rescue Squad 21
          Stations:   2, 4, 12, 15, 17, 25, 31, 34, 36,
                    and  38

IV.  EQUIPMENT

A. Safety Regulation

Technical rescue equipment shall be inventoried and
inspected in accordance with SOP 2.3.04, Personnel Safety
(XII. Daily Inspections) and verified by signing FSA-41
as required in SOP 3.6.02, Vehicle Maintenance Log.

B. Technical Rescue Equipment Inventory

1.  Technical Rescue Squad

a.  Technical Rescue Squads shall carry the following
equipment:

1)  Standard Rescue Squad Inventory
2)  Rope Rescue equipment specified in the Rope
Operations Manual
3)  Technical Rescue Equipment

a)  Technical Rescue ICS Charts, Vests &
Communications System and Personnel
Identification Arm-Bands (6-Orange)
b)  Air Bags, Low Pressure
Atmospheric Monitors
c)  Breathing Apparatus - Supplied Air System
d)  Concrete Drill & Breaker (Pionjar-TM)
f)  Ejector (16") with    25' Extension  Tube
Hilti Gun (TM)
g)  Jimmi Jaks (TM)
i)  Light Sticks (Caylume-TM)
j)  Lighting System, Low-Voltage, Explosion
Proof
k)  Pak-hammer 90 (TM)

2.  Technical Rescue Unit

a. Technical Rescue Units shall maintain the
following minimum equipment inventory:

| Q# | UNIT | SIZE | DESCRIPTION |
|---|---|---|---|
| 9 | e | 18-27" | AIRSHORES - A (BLACK) |
| 9 | e | 23-37" | AIRSHORES - B (RED) |
| 9 | e | 30-48" | AIRSHORES - C (WHITE) |
| 9 | e | 42-66' | AIRSHORES - D (ORANGE) |
| 9 | e | 60-96" | AIRSHORES - E (YELLOW) |
| 9 | e | 90-120' | AIRSHORES - F (BLUE) |
| 9 | e | 114-144" | AIRSHORES - G (GREEN) |
| 6 | e | ........ | APRONS,  CARPENTER'S |
| 4 | e | 14" | BARS, WONDER (3-3/4" X 13-23/8") |

| Q# | UNIT | SIZE | DESCRIPTION |
|---|---|---|---|
| 2 | @ | 18" | BARS, WRECKING |
| 2 | @ | 30" | BARS, WRECKING |
| 2 | @ | 60" | BARS, PINCH |
| 6 | @ | 5 GAL. | BUCKETS, PLASTIC |
| 2 | @ | 100' | CHALK LINES, RED/BLUE |
| 4 | CYL. | 20 LB. | CO2 W/ 2-CYLINDER CART, REGULATOR & HOSE |
| 1 | 100' | . . . . .CO2 HOSE (RED) W/ REGULATOR (SPARE) | |
| 3 | @ | . . . . . .CRATES, PLASTIC (STORAGE) | |
| 6 | @ | . . . . . . . GLASSES, SAFETY | |
| 6 | @ | 4x8 | GROUND PADS |
| 2 | @ | 8 LB. | HAMMERS, SLEDGE |
| 2 | @ | . . . . . . . . HAMMERS, ENGINEER'S, SHORT HANDLE | |
| 4 | @ | . . . . . .HAMMERS, CLAW | |
| 6 | @ | . . . . . .HARD HATS, SAFETY | |
| 4 | @ | . . . . . . KNIFE, UTILITY | |
| 2 | @ | 16' | LADDERS, STRAIGHT |
| 1 | BOX | 50 LB& | NAILS, DOUBLE-HEAD |
| 2 | @ | | PAINT, SPRAY (BLUE..:WATER) |
| 2 | @ | | PAINT, SPRAY (GREEN.:SEWER) |
| 2 | @ | | PAINT, SPRAY (ORANGE:COMMUNICATIONS) |
| 2 | @ | | PAINT, SPRAY (RED...:ELECTRIC) |
| 2 | @ | | PAINT, SPRAY (WHITE.:EXCAVATION) |
| 2 | @ | | PAINT, SPRAY (YELLOW:GAS/OIL) |
| 1 | @ | 50 GPM | PUMP, ELECTRIC, SUBMERSIBLE w/ 50' - 1 & 1/2" HOSE |
| 1 | @ | 250 GPM | PUMP, TRASH (2) 10' HARD SLEEVES (1) 25' 3" HOSE |
| 8 | 25' | 1/2" | ROPE, NYLON |
| 2 | 12' | 1/2" | ROPES W/ WIRE SLINGS |
| 2 | @ | 6' | RULERS, FOLDING |
| 2 | @ | 26" | SAW, HAND, 8PT |
| 20 | @ | 2x4xXx | SCABS |
| 12 | @ | 4x8 | SHORING PANELS (17-PLY) W/ 2x12 UPRIGHTS (BOLTED) |
| 6 | @ | . . . . . . SHOVELS, ENTRENCHING | |
| 2 | @ | . . . . .SHOVELS, ROUND POINT, LONG HANDLE | |
| 4 | @ | 24" | SQUARE, CARPENTER'S |
| 24 | @ | . . . . . SWIVELS, AIRSHORE | |
| 2 | @ | 12' | TAPE MEASURES |
| 4 | @ | 30' | TAPE MEASURES |
| 2 | @ | 100' | TAPEMEASURE, FIBERGLASS |
| 2 | @ | 12x16' | TARPS, VINYL |
| 2 | @ | 4X4X8 | TIMBERS |
| 4 | @ | 6X6X8 | TIMBERS |
| | @ | | TRIPOD, (ROLLGLISS-TM) |
| 1 | @ | . . . . . TOOL BOX | |
| 2 | @ | 2X12X8 | UPRIGHTS |
| 2 | @ | 2X12x12 | UPRIGHTS |
| 12 | @ | . . . . . WEDGES, WOOD | |

## V. OPERATIONS

### A. Scene Management

1. Responsibility

   Department personnel are responsible for compliance with the provisions of Chapter V., Part A Scene Management on all Technical Rescue Box Alarm Assignments and training evolutions.

2. Command

   a. Technical rescue operations shall be conducted in accordance with the Incident Command System Manual and the Technical Rescue Incident Command System.

   b. The officer in charge of the first-arriving fire department unit shall establish command, issue a situation report and direct the crew to perform scene control operations.

      1) The situation report shall include a response recommendation for the TROT based oh the degree of difficulty involved in the rescue and the capabilities and limitations of the alarm assignment.

   C. The responding battalion chief shall determine if the TROT is required.

      1) Technical communications, prior to the arrival of technical rescue apparatus, shall be directed to the first-due technical rescue squad.

      2) The officer in charge of the first-due technical rescue squad say recommend, to the responding battalion chief, ah increased technical rescue alarm assignment based on dispatch information or the situation report.

   d. The officer in charge of the first-arriving technical rescue apparatus shall assume the position of technical rescue leader.

      1) The technical rescue leader shall consult with the officer of the first-arriving fire department unit and complete the incident size-up while the crew assists with scene control operations.

3.  Scene Control

    a.  Department personnel are prohibited from
        entering any hazardous area without the proper
        protective equipment and until the area has
        been rendered safe through appropriate special
        teem actions. Examples of these hazardous
        areas us:

        1)  Confined Space
        2)   Hazardous Material Environments
        3)  Structural Collapse Voids
        4)  Trench (Depth > 5')

    b.  Scene Control shall be established as follows:

        1)  Position apparatus appropriately, assess
            and mitigate immediate life-threatening
            hazards and provide medical and protective
            support to accessible victims. Incident
            specific scene control measures are listed
            in Chapter V., Part C.

        2)  Establish a 100' diameter rescue site
            area, or appropriately sized haz-mat
            zones, post the area with Fire-Line Tape
            (TM), and remove all non-essential
            civilian and department personnel.

        3)  Establish the equipment and personnel
            areas adjacent to the technical rescue
            unit.

    c.  Department personnel shell report to the
        rescue personnel officer/area upon completion
        of initial scene control activities, and stand
        by for the teas briefing.

4.  Teas Briefing

    a.  A team briefing shall be conducted, following
        the completion of initial scene control
        operations, prior to the commencement of
        rescue operations.

        1)  The senior (TRO-II) officer shall assume
            the position of technical rescue leader,
            confer with the incident commander, assign
            the Technical Rescue Incident Command
            System (TR ICS) positions and review the
            Incident Action Plan with all personnel.

a) The following positions shall be
staffed on all incidents and the
personnel shall be identified by lime-
yellow commend vests. The duties end
responsibilities of each position are
listed in Chapter V., Put B. Tactical
- Chain of Command.

(i)   Technical Rescue Leader
(ii)  Rescue Sector (Optional)
(iii) Rescue Safety
(iv)  Rescue Equipment
(v)   Rescue Personnel

b) The rescue site shall be sketched on
the ICS board at the technical rescue
unit and the Incident Action Plan will
be reviewed with all fire department
personnel.

( i)  Officers shall utilize the
Technical Rescue ICS Chart to
assist them in managing the
incident.
(ii)  Deviation from the established
Incident Action Plan is
prohibited without authorization
from the incident commander.

2) Personnel shall remain in the personnel
area until assigned a task by the rescue
personnel officer. Personnel shall report
back to the rescue personnel officer upon
completion of the task.

## B. Tactical - Chain of Command

1. Battalion Chief (Box Running Order)

2. Technical Rescue Leader (Certified TRO-II Officer)

a. The technical rescue leader is responsible for
the tactical operation of  the team under the
direction of the incident commander or
operations officer.

1) The technical rescue leader will normally operate at the rescue site providing direct supervision of the operation.

2) On sectorized incidents, the technical rescue leader will operate at the technical rescue unit, command post or other area, as designated by the incident commander or operations officer, and be responsible for the management of the technical rescue component of the incident through rescue sector officers.

b. The technical rescue leader is responsible for submitting a written report, detailing the team's activities, to the senior technical officer and the TSO within seven days of an incident.

3. Rescue Sector (Certified TRO-II Officer)

Rescue sector officers are responsible for the direct supervision of tactical operations of the team, in a sector, under the direction of the technical rescue leader.

4. Rescue Safety (Certified TRO-II Officer)

a. The rescue safety officer is responsible for the safety of the team in conjunction with the technical rescue leader and the incident commander.

1) The rescue safety officer will normally operate at the rescue site, in a physical position opposite the technical rescue leader, assisting with the direct supervision of the operation.

2) On sectorized incidents, a rescue safety officer will be assigned to each sector. The rescue safety officer will operate at a sectorized rescue site, in a physical position opposite the rescue sector officer, assisting with the direct supervision of the operation.

b. The rescue safety officer is authorized and required to stop and correct any unsafe operation.

5.    Rescue Personnel (TRO-II Personnel)

   a.   The rescue personnel officer is responsible for managing the personnel area at the technical rescue unit and controlling access to the rescue site(s).

   b.   The rescue personnel officer shall issue armbands to personnel authorized to operate in the rescue site area.

6.    Rescue Equipment (TRO-II Personnel)

   a.   The rescue equipment officer is responsible for managing the rescue equipment and resources in the equipment area at the technical rescue unit.

   b.   The rescue equipment officer may obtain emergency supplies through the duty Resource Management officer.

7.    Technical Rescue ICS Chart

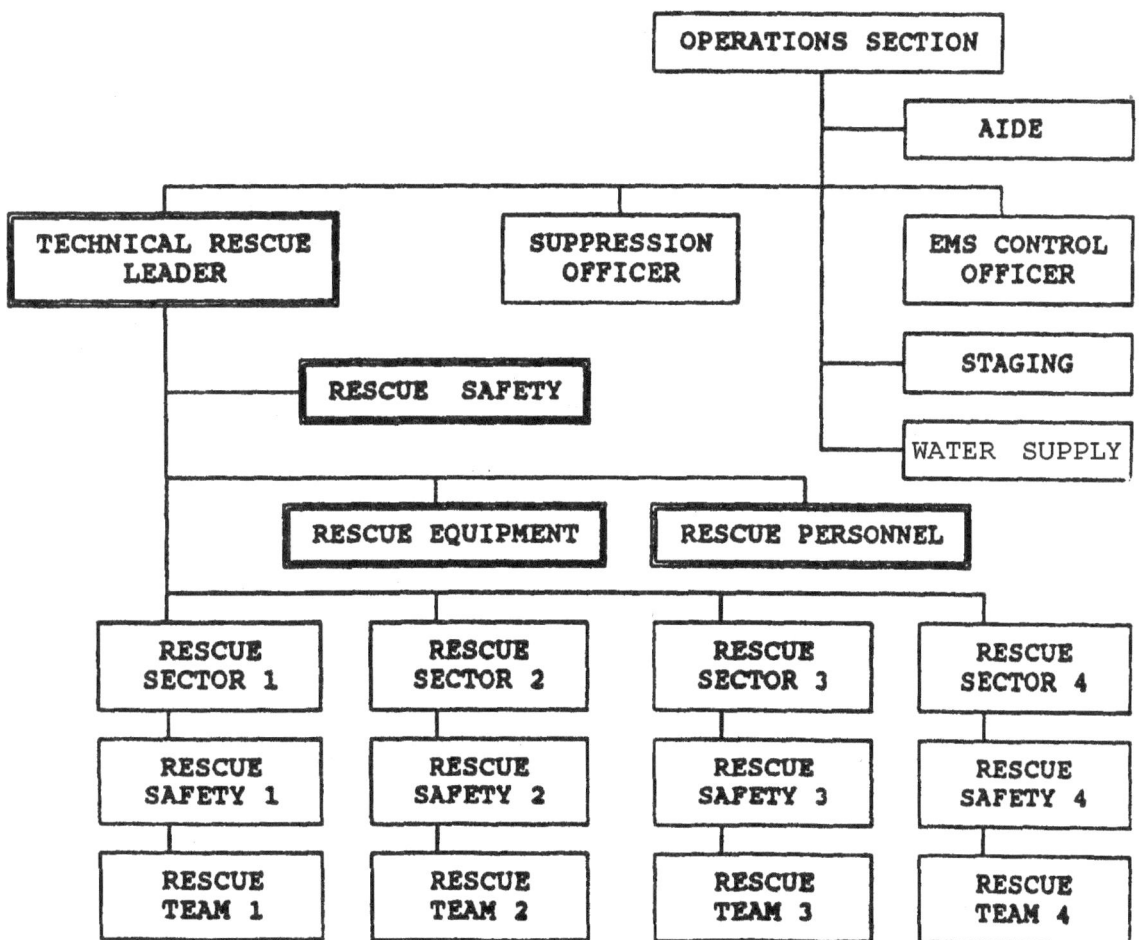

C. Technical Rescue Operations

    1. Confined Space Rescue

        a. Confined space rescue operations shall be governed by the provisions of this manual and the procedures established in the TROT Training Outline: Confined Space Rescue Operations.

        b. Statutory Regulation - Commonwealth of Virginia VR 425-02-12 Virginia Operational Safety & Health Standards for General Industry - Virginia Confined Space Standard 1910.146. (Statutory Authority: 40.1-22 of the Code of Virginia. Effective: July 1, 1987.)

            1) VR 425-02-12 : Section 1. Definitions.
Confined Space. A confined space is any space not intended for continuous employee occupancy, having a limited means of egress, and which is also subject to the accumulation of an actual or potentially hazardous atmosphere or a potential for engulfment.
Examples: Storage tanks, process vessels, bins, boilers, ventilation and exhaust ducts, sewers, manholes, underground vaults, acid tanks, digesters, ovens, pulpers, tunnels, pipelines and open top vessels more than four feet in depth such as pits, trenches, tubs, vaults and vessels.
Rescue Team: Rescue team means those persons whom the employer has designated prior to any confined space entry to perform rescues from confined spaces. The rescue team may consist of outside personnel provided the training requirements of part [7. A. 2.] of this standard are met.

          2) VR 425-02-12 : Section 7. Training., (A. 2.]
[A. The employer shall inform his employees of the hazards of working in confined spaces by providing specific training to employees before they may be authorized to enter a confined space.
    2. Rescue teams. Rescue teams shall be trained to use the equipment they may need to perform rescue functions assigned to them.

a.   At least annually rescue teams shall practice removing victims through openings and portals of the same size, configuration and accessibility as those of spaces from which an actual rescue could be required.

b.   The attendant or at least one member of each rescue team shall hold current certification in basic first aid and CPR (Cardio-Pulmonary Resuscitation).]

c.   Confined Space Entry Permit

The technical rescue leader or rescue sector officer shall not authorize a confined space entry, during training or incident operations, until the rescue safety officer issues a Fairfax County Fire & Rescue Department - Technical Rescue Team - Confined Space Entry Permit (FSA-237).

2.   Rope Rescue

Rope rescue operations shall be governed by the provisions of this manual and the procedures established in the Department Operating Manual: Rope Operations.

3.   Structural Collapse Rescue

a.   Structural collapse rescue operations shall be governed by the provisions of this manual and the procedures established in the Technical Rescue Team Training Outline: Structural Collapse Rescue Operations.

b.   The TROT shall operate in collapsed structures for the sole purpose of conducting search and rescue operations.

4.   Trench Rescue

Trench rescue operations shall be governed by the provisions of this manual and the procedures established in the Technical Rescue Team Training Outline: Trench Rescue Operations.

# TECHNICAL RESCUE OPERATIONS CHECKLIST

## CONFINED SPACE RESCUE
### ROPE RESCUE
#### STRUCTURAL COLLAPSE RESCUE
##### TRENCH RESCUE

### SIZE-UP

| | | | | |
|---|---|---|---|---|
| [ ] | [ ] | [ ] | [ ] | **RESEARCH** |
| [ ] | [ ] | [ ] | [ ] | Alarm Information / Supervisor / Witnesses |
| [ ] | | | | Maps / Pre-Plans / Blue Prints / Cut Sheets |
| | | | | Confined Space Entry Permit |
| | | | | * Contents / Storage / Operations |
| | | | | * Atmosphere / Job Description |
| | | | | **SITE CONDITIONS** |
| [ ] | [ ] | [ ] | [ ] | Terrain / Elevation / Positioning / Staging |
| [ ] | [ ] | [ ] | [ ] | Dimensions |
| | | [ ] | [ ] | Collapse Type / Scope / Occupancy |
| | | | [ ] | Soil Type / Ground Water / Dewatering System |
| | | | | Tension Cracks / Overhangs / Subsidence |
| | | | | Bulging / Spoil Pile / Superimposed Loads |
| [ ] | [ ] | [ ] | [ ] | Utilities: Compromised / Damaged / Exposed |
| | | | | **VICTIM SITUATION** |
| [ ] | [ ] | [ ] | [ ] | Elapsed Time |
| [ ] | [ ] | [ ] | [ ] | Number of Victims / Assign ID Numbers |
| | | | | Location / Condition / Entrapment of @ |
| | | | | **SPECIAL ALARMS** |
| [ ] | [ ] | [ ] | [ ] | Personnel / Equipment / Supplies |

### SCENE CONTROL / SAFETY TASKS

| | | | | |
|---|---|---|---|---|
| [ ] | [ ] | [ ] | [ ] | Clear Area / 100' Rescue Site Area |
| [ ] | [ ] | [ ] | [ ] | Scene Lighting |
| | | [ ] | [ ] | Collapse Zone / Stability Monitors |
| | | | [ ] | Ground Pads |
| | | | | Victim Aid |
| [ ] | | [ ] | [ ] | * Supplied Air (Low) / Helmet |
| | [ ] | | | * Rapid Access / Tag Line / EMS Exam / Helmet |
| [ ] | | [ ] | [ ] | Atmospheric Monitoring / Ventilation Systems |
| [ ] | | [ ] | [ ] | Dewatering System |
| [ ] | [ ] | [ ] | [ ] | Utilities / Machinery |
| | | | | * Remove Superimposed Loads / Secure & Support |
| | | | | * Lock-Out & Tag / Disconnect & Blind |
| | | | | * Block @ Zero Mechanical State (ZMS) |
| | | | | * Test Check Safety Systems |
| [ ] | | [ ] | [ ] | 500' Vibration Zone: Traffic / Rail / Blasting |

```
┌─────────────────────────────────────────────────────────────────┐
│              TECHNICAL RESCUE OPERATIONS CHECKLIST                │
└─────────────────────────────────────────────────────────────────┘
```

## CONFINED SPACE RESCUE

### ROPE RESCUE

#### STRUCTURAL COLLAPSE RESCUE

##### TRENCH RESCUE

###### RESCUE OPERATIONS

| | | | | |
|---|---|---|---|---|
| | | | | **TEAM BRIEFING** |
| [ ] | [ ] | [ ] | [ ] | ICS Staff / Sectors / Communications System |
| [ ] | [ ] | [ ] | [ ] | Rescue Teams 1 & 2 / Action Plan |
| | | | | **SAFETY OFFICER** |
| [ ] | | [ ] | [ ] | Atmospheric Monitoring / Ventilation Systems |
| [ ] | | [ ] | [ ] | FCFD Confined Space Entry Permit |
| [ ] | [ ] | [ ] | [ ] | Rope Rescue System Checks |
| [ ] | [ ] | [ ] | [ ] | Personnel Rotation |
| | | | | **OPERATIONS** |
| | | [ ] | [ ] | Air Shoring System Pressure |
| | | | | * Structural Collapse = 50 lbs. Maximum. |
| | | | | * Trench ............. = 200 lbs. |
| | | | | Trench Shoring Order...:  1ST SET      2ND SET |
| | | | | * Air................. =  M-T-B        M-B-T |
| | | | | * Timber & Pipe....... =  T-M-B        M-B-T |
| | | | | **VICTIM RESCUE** |
| [ ] | [ ] | [ ] | [ ] | Extrication / Exam / Packaging / Removal |

###### SUPPORT / NOTIFICATIONS

| | | | | |
|---|---|---|---|---|
| [ ] | [ ] | [ ] | [ ] | Command / Communications Pod |
| [ ] | [ ] | [ ] | [ ] | Light & Air Unit |
| [ ] | [ ] | [ ] | [ ] | Hazardous Materials Unit |
| [ ] | [ ] | [ ] | [ ] | Dive Rescue Team |
| [ ] | [ ] | [ ] | [ ] | Environmental Management |
| [ ] | [ ] | [ ] | [ ] | Labor & Industry |
| [ ] | [ ] | [ ] | [ ] | Utilities |
| | | | | * Gas / Pipeline / Power / Telephone / Water |

| UNDERGROUND FACILITY PROTECTION ACT CODE of VIRGINIA, Chapter: 10.3, 56-265 | | |
|---|---|---|
| COLOR | UTILITY | SYMBOL |
| BLUE..........WATER.........: | | -W- |
| GREEN.........SEWER.........: | | -S- |
| ORANGE........COMMUNICATIONS: | | -TV-/-TEL- |
| RED...........ELECTRIC......: | | -E- |
| WHITE.........EXCAVATION....: | | |
| YELLOW........GAS/OIL.......: | | -G-/-O- |

| MAXIMUM DEPTH = 15' | SLOUGH-IN ... = 18" |
|---|---|
| WALER    0'-7' = 4x4 | HORIZONTAL .. =    4' |
| WALER    7' +  = 6x6 | VERTICAL .... =    4' |
| HEAVY R/C  = 150/CF<br>LIGHT R/C  = 100/CF<br>  4" SLAB   =  50/SF<br>  6" SLAB   =  75/SF<br>  8" SLAB   = 100/SF<br>  2'x16"x10'= 4000 #<br> 10'x10'x6" = 7500 # | BLACK .... 18 - 27"<br>RED ...... 23 - 37"<br>WHITE .... 30 - 48"<br>ORANGE ... 42 - 66"<br>YELLOW ... 60 - 96"<br>BLUE ..... 90 -120"<br>GREEN ....114 -144" |

### SHORING FOR HARD COMPACT SOIL

| DEPTH (feet) | Uprights Horizontal spacing (feet) | Uprights Size (inches) | Stringer (Waler) Size (inches) | Wood size (inches) & excav widths (feet) | Aluminum Pipe & Hydraulic Sys Min Dia (in)° | Aluminum Max excav width (ft) | Steel Pipe & Hydraulic Sys Min Dia (in) | Steel Max Excav (ft) width |
|---|---|---|---|---|---|---|---|---|
| 5 to 7 | 8 | 3 x 6 | — | 4 x 4 all widths up to 15' | 2½ (3½) | 8 (10) | 1½ | 3 |
|  | 4 | 2 x 10 | 4 x 4 |  | 2½ (3½) | 8 (14) | 1½ | 3 |
|  | 2 | 2 x 8 | 4 x 4 |  | 2½ (3½) | 8 (20) | 1½ | 3 |
| over 7 to 10 | 8 | 4 x 10 | — | 4 x 4 up to 12' width, over 12' up to 15' 6 x 6 | 2½ (3½) | 6 (8) | 2 | 6 |
|  | 4 | 3 x 10 | 6 x 8 |  | 2½ (3½) | 9 (11) | 2½ | 12 |
|  | 2 | 3 x 8 | -6 x 8 |  | 2½ (3½) | 12 (16) | 3 | 15 |
| over 10 to 12 |  |  |  | 4 x 4 up to 8' width, over 8' up to 15' 6 x 6 | 2½ | 6 | 2 | 8 |
|  | 8 | 6 x 8 | — |  | 3½ | 7 | 2½ | 12 |
|  |  |  |  |  | 2½ | 8 | 2 | 10 |
|  | 4 | 4 x 6 | 8 x 8 |  | 3½ | 10 | 2½ | 11 |
|  |  |  |  |  | 2½ | 10 | 2½ | 13 |
|  | 2 | 3 x 6 | 8 x 8 |  | 3½ | 15 | 3 | 15 |
| over 12 to 15 | 8 | 6 x 8 | — | 4 x 4 up to 6' width, over 6' up to 15', 6 x 6 | 2½ | 5 | 2 | 6 |
|  |  |  |  |  | 3½ | 6 | 2½ | 10 |
|  |  |  |  |  | 2½ | 7 | 2 | 8 |
|  | 4 | 4 x 10 | 8 x 10 |  | 3½ | 9 | 2½ | 12 |
|  |  |  |  |  | 2½ | 9 | 2½ | 13 |
|  | 2 | 3 x 10 | 8 x 10 |  | 3½ | 13 | 3 | 15 |
| Over 15 to 20 | 8 | 6 x 10 | — | 6 x 6 up to 14' width over 14' up to 20' 8 x 8 | 2½ | 4 | 2½ | 8 |
|  |  |  |  |  | 3½ | 5 | 3 | 12 |
|  | 4 | 4 x 12 | 6 x 12 |  | 2½ | 6 | 2½ | 10 |
|  |  |  |  |  | 3½ | 8 | 3 | 15 |
|  |  |  |  |  | 2½ | 8 | 2½ | 12 |
|  | 2 | 3 x 12 | 6 x 12 |  | 3½ | 11 | 3 | 15 |
| Over 20 | See Section 1541 (a) (6) |  |  |  |  |  |  |  |

# FAIRFAX COUNTY
## FIRE AND RESCUE DEPARTMENT
## TECHNICAL RESCUE OPERATIONS TEAM

### CONFINED SPACE ENTRY PERMIT

| DATE: - - | TIME: : | BOX# | I# |
|---|---|---|---|

**ADDRESS:**

**SITUATION:** [ ] RESCUE  [ ] SEARCH  [ ] RECOVERY  [ ] TRAINING

TYPE OF SPACE.....:

NORMAL OPERATIONS.:

NORMAL STORAGE....:

### ATMOSPHERIC MONITORING

| TIME | | | | | | | | |
|---|---|---|---|---|---|---|---|---|
| OXYGEN | | | | | | | | |
| TOXINS | | | | | | | | |
| GASES | | | | | | | | |

### SCENE CONTROL / SAFETY TASKS

| | |
|---|---|
| ___ | CHECK ON-SITE ENTRY PERMIT |
| ___ | VENTILATION SYSTEMS |
| ___ | LOCK-OUT : ELECTRICAL |
| ___ | LOCK-OUT : MECHANICAL |
| ___ | LOCK-OUT : PNEUMATICS |
| ___ | LOCK-OUT : AUXILIARY |
| ___ | ZERO MECHANICAL STATE |
| ___ | AIR SUPPLY |
| ___ | BACK-UP AIR SUPPLY |
| ___ | ENGULFMENT POTENTIAL |
| ___ | TAG LINES |
| ___ | INDEPENDENT SUPPORT LINE |
| ___ | LIGHTING |
| ___ | COMMUNICATIONS SYSTEM |
| ___ | VESSEL PURGE (RECOVERY) |
| ___ | BACK-UP TEAM READY |

### RESCUE TEAM

| ENTRY | TEAM #1 | EXIT |
|---|---|---|
| | | |
| | | |
| | | |
| | | |

| ENTRY | TEAM #2 | EXIT |
|---|---|---|
| | | |
| | | |
| | | |
| | | |

### SAFETY OFFICER

FSA-237 (07/91)

This Confined Space Entry Permit is required by Section 6. [A] Permit Systems, of the Virginia Confined Space Standard-1910.146.VR 425-02-12. Virginia Occupational Safety and Health Standards for General Industry. Statutory Authority: 40.1-22 (5) of the Code of Virginia.

6. [A] [l.] The minimum acceptable environmental conditions for entry and work in a Confined Space are specified by Section 3., Preparation., of the Virginia Confined Space Standard and referenced in the Department Operating Manual (DOM) - Technical Rescue.
a. Pumps or lines which may convey flammable, injurious or' incapacitating substances into a space shall be disconnected, blinded or effectively isolated by other means to prevent the development of dangerous levels of air contamination or oxygen deficiency within the space. The disconnection or blind shall be so located or done in such a manner that inadvertent reconnection of the line or removal of the blind are effectively prevented.
b. All fixed mechanical devices and equipment that is capable of causing injury shall be placed at zero mechanical stated (ZMS).
c. Electrical equipment, excluding lighting shall be locked out in the open (off) position with a key-type padlock. The key shall remain with the entering Rescue Team. Where it is impossible to accomplish the lock-out with a padlock, the equipment shall be tagged in accordance with 1910.145 (f) and the Firefighter shall be stationed by the switch(s) for the duration of the incident.
d. Confined spaces shall be emptied, flushed or otherwise purged of flammable, injurious, or incapacitating substances to the extent feasible.
e. When a hazardous atmosphere exists the space shall be mechanically ventilated until the concentration of the hazardous substance(s) is/are reduced to a safe level and continued for the duration of the incident.
6. [A] (2.) Atmospheric testing shall be conducted prior to entry and the results entered in the first time block under Atmospheric Monitoring on the reverse side of this form. Subsequent monitoring shall continue every 15 minutes for the duration of the incident.
6. [A] [3.] Calibration of testing instruments shall be performed annually or as required by Hazardous Materials Technicians at Fire Station 34. Testing instruments shall be field checked prior to use.

| INSTRUMENT | PROPERTY # | CALIBRATION DATE |
|---|---|---|
| Oxygen Detector | | |
| Combustible Gas Detector | | |

6. [A] (4.) The Safety Officer is the "qualified person" responsible for completing and issuing this permit prior to entry.
6. [A] (5.) A written description of location and scope of work shall appear in the top block on the reverse side of this form.
6. [A] [6.] This permit expires 12 hours after the incident time recorded in the top block on the reverse side of this form.
6. (B) This form shall be attached to the Technical Rescue Leader's written report to the Senior Technical Officer. (DOM: V.,B.,2.,b.)

# METROPOLITAN DADE COUNTY FIRE RESCUE DEPARTMENT
## WATER RESCUE BUREAU
## WATER RESCUE POLICY AND PROCEDURE

**REVISED DRAFT   10-92**

## 26.01    PURPOSE:
To provide appropriate guidelines for the conduct of Fire Department Water Rescue Operations.

## 26.02  POLICY:
In order to assure safe operations at Water Rescue incidents, all Fire Department personnel shall respond and conduct themselves in the manner outlined in this policy.

## 26.03    AUTHORITY
The authority vested in the Fire Chief by Florida Statute 125.01., Sections 4.01 and 4.02 of the Metro Dade County Charter, Section 2-18 of the Code of Metropolitan Dade County.

## 26.04    RESPONSIBITY:
It shall be the responsibility of all personnel to thoroughly familiarize themselves with and conform to this policy.  It shall be the responsibility of all Fire Department supervisors and officers to supervise and command their subordinates in conformance with this policy. It shall be the responsibility of the Water Rescue Bureau  to update and revise this policy as necessary.

## 26.05    PROCEDURE
The Communications Division will maintain a Master Roster of Certified Rescue Skin Divers, Department SCUBA Rescue Authorized Divers and SCUBA Rescue Authorized Divers Suffix "A" (Air Delivery Qualified).

Prior to station recall check each morning, the Unit OIC will review personnel in his/her unit to determine availability of approved water rescue personnel. The Unit OIC will advise the Alarm Office during morning recall check of the following information:

1.    Identify Unit: Suppression, Rescue, Battalion and specialty units: Air Truck, Hazmat, Water Rescue, etc.

2.    Number of SCUBA Rescue Authorized Diver(s) per unit and whether "A-Qualified" (Air Delivery Qualified) or not ("B-Diver). Example: "Rescue 3, one A Diver, 2 B divers).

The Unit OIC will advise the Alarm Office of any changes in capability status during the shift.

Based on information received from all Unit OIC's, the Alarm Office will maintain a daily roster of units which have SCUBA Rescue Authorized Diver(s), the number of divers per unit, and whether "A Diver(s) " or "B Diver(s).

The Water Rescue Bureau will maintain a list of diver I.D. numbers.

After notification to the Alarm Office, the Unit OIC, in agreement with the Senior Rescue Diver, will designate:   personnel assignments in the following  categories:

1.     Primary Rescue Diver

      a.     Department SCUBA Rescue Authorized Diver
      b:     Department Certified Rescue Skin Diver

2.     Safety Diver

      a.     Department SCUBA Rescue Authorized Diver
      b.     Department Certified Rescue Skin Diver

Since involvement in Rescue Diving is voluntary and not a job requirement, Rescue Divers may call "No Dive". It is the responsibility of the OIC to respect this decision.

Following is a list of reasons for calling "No Dive". This list is not to be considered complete but is presented for guidance only:

Safety Reasons
      Skill level inadequate for environmental conditions
      Dive buddy or safety diver inadequate or not present
      Dive rescue equipment malfunction
      Senior Rescue Diver judgement call per SOP

Medical Reasons:
      Ear, nose; throat congestion or blockage
      Respiratory infection
      Dehydration and/or fatigue
      Phobia
      Senior Rescue Diver judgement call per SOP

3.    Line Tender

    a.    Shall be familiar with Water Rescue Line Signals
    b.    Shall understand Water Rescue Search Pattern Procedures

Each morning all dive rescue equipment shall be checked as follows:

1.    Basic Water Rescue 'equipment will be checked out by department certified rescue skin or SCUBA Authorized divers.

2.    SCUBA Rescue equipment will be checked out by SCUBA Rescue Authorized divers.

3.    When a Department SCUBA Rescue Authorized Diver is not assigned to a unit, the Unit OIC shall verify that all dive rescue equipment is accounted for and secure, It is expected that fire units without specialized water rescue capacity will arrive on the scene prior to units with certified dive rescue personnel. The OIC should manage the scene utilizing the individual firefighter's skills just as they would be utilized to perform best effort available at other type of incidents. Utilize this time to gather information: datum point, number of patients, patients' clothing, patients age, time frame and witnesses., in anticipation of the arrival of Rescue Skin Divers or SCUBA Rescue Authorized Divers. Incident Command Procedures will be established.

Safety of the firefighter is a primary concern. Therefore, the use of self contained breathing apparatus (SCBA) during water rescue incidents, and the use of SCUBA by unauthorized personnel, is strictly prohibited.

Initial dispatch on a water rescue incident shall consist of the following:

1.    Closest Unit (Suppression, Rescue, Battalion, or Specialty Unit).

2.    Closest Unit with SCUBA Rescue Authorized Divers, A MINIMUM OF TWO SCUBA AUTHORIZED DIVERS WILL BE DISPATCHED!!!

3.    Rescue Unit - If primary divers are on this rescue unit, a second rescue unit will be dispatched.

4.    Battalion Commander

In addition, all Water Rescue Bureau personnel will be advised of the incident by pager.

The Battalion Commander, as the Incident Commander, shall be responsible for surface control of the water rescue scene - responsibilities include:

1.  Secure the scene (safety of personnel & bystanders)

2.  Crowd Management (request of Metro Dade Police Department).

3.  Keep a roster of personnel entering the water: time in, time out and psi used.

4.  Resource Management

    Request of additional Fire or Rescue units ( For Manpower)
    6:  Request of Support personnel and/or equipment ie... Air rescue, Hazmat, Air truck, scene support and or Water Rescue Staff units.

5.  Precluding personnel from entering water who are not requested by the senior rescue diver.

6.  Designating personnel to prepare rescue unit for immediate patient care (set up airway management, (Ambu, suction, 02 & E.T.equipment} towels, defibrillator, IV and cardiac medication therapy, backboard, thumper and stretcher.)

7.  Designating personnel to prepare for removal of patient(s) from water ie. backboard, stokes stretcher, telescoping latter, roof latter, rope, boat and or Air rescue hoist.

The responsibility of Water Rescue Operations shall be assigned to the Senior Rescue Diver on the scene (water sector) to insure that all water rescue activities adhere to all Fire Department standard operating procedures. ( This is based on qualifications and experience not rank, seniority, or date of authorization.) It shall be the responsibility of the Senior Rescue Diver to determine all unsafe conditions and to cancel operations, if found to be life threatening. Incident command procedures should be used to insure that effective and accurate information is being deciminated. Command must be identified by radio transmission.

If a determination has been made, by the Battalion Commander and the Senior Rescue Diver, that the search and rescue operation is no longer a rescue, but a body recovery, all Fire Department diving operations shall be suspended and control of the accident scene shall be transferred to MDPD Divers for body recovery.

Note: If MDPD Divers are unable to respond, Fire Department Divers may continue the search at the discretion of the Battalion Commander and the Senior Rescue Diver. A complete break, minimum of 15 minutes, scene reassessment and revised dive plan is recommended before divers who were involved in a "rescue effort" direct their energies to body recovery. This will insure time-stress factors are not carried over to body recovery mode.

In an effort to maintain our objective to provide an emergency search and rescue service, all requests from outside agencies for special services shall be made thru the Alarm Office by that agency's ranking representative The request shall include the following:

1. Location
2. Reason for request
3. Type of object to be recovered or service to be rendered

Upon approval, a Department SCUBA Rescue Authorized diver(s) shall be assigned to the operation. If the request is for evidence recovery, all efforts shall be made to safeguard the integrity of the evidence and scene.

If the request is for ship-bottom survey, all water induction and propulsion systems shall be shut down and tagged for safety. A FD officer shall be assigned to the bridge to insure "shut down" until diving operations are completed.

All mandatory dive equipment shall be employed.

For Water Rescue Operating Procedure see "MDFR Water Rescue Resource Manual".

# WATER RESCUE OPERATING PROCEDURES FOR RESCUE SKIN DIVER

## I. Rescue of an active drowning victim (surface recovery).

Note:       On all night operations, all personal flotation (BC or PFD) shall have a cyalume light stick attached and activated.

A.    Encourage, Reach, Throw, Row and Go (Water Entry, "Go" is the m os t dangerous rescue option and may be least appropriate.

B.    Rescue Skin Diver (with mandatory and accessory equipment, see Attachment #1)

    1.    Keep victim in sight at all times
    2.    Make proper water entry
    3.    Use Head High approach
    4.    Employ appropriate Water Rescue Evolution
        a.    Torpedo Buoy
        b:    Line Rescue
        C.    Unassisted Rescue (least desirable option)

C.    Safety Diver (Wearing mandatory and accessory equipment see Attachment #1)

    1.    Keep victim and Rescue Diver in sight at all times
    2.    Note victim's location using geographic references
    3.    Render assistance to Rescue Diver as necessary

D    Line Tender (with BC & foot protection), controls surface directed search.

    1.    Keep victim and Rescue Diver in sight at all times
    2.    Note victim's location using geographic references
    3.    Prepare to help victim ashore
    4.    Monitor hand-held radio

Note:       After recovery of a viable drowning victim conduct appropriate in-water medical treatment as indicated. Remove patient from water with assistance of shore personnel. Use adjuncts when possible (roof ladders, aerials, etc.). Treat patient per appropriate EMS protocol.

## II. Search and Rescue of a drowning victim - Surface Directed from Shore:

Note:        On all night operations, all personal flotation (BC or PFD) shall have a cyalume light stick attached and activated.

A.    Rescue Skin Diver (with mandatory and accessory equipment, see Attachment #1)

    1.    Determine Datum Point
    2.    Establish Area of Probability
    3.    Confirm type of Search Pattern

Note:        If a known Datum Point is available a Circular Pattern should be implemented. The use of a Linear Pattern should be employed when a Datum Point is not known.

    4.    Confirm Rescue Line Signals with Line Tenders: OATHS

        a.    One Pull =        OK
        b:    Two Pulls =        Advance Line (Give Me Line)
              Three Pulls =      Take up Line (Pull Me In)
        d:    Four Pulls =       Help! (Secure Line & Assist Me) (1)
        e.    Five Pulls =

            (1)    If Help signal is received, tie loop in line to mark diver distance, note direction to diver; alert safety diver to respond.

    5.    Rig torpedo buoy on water rescue line (see Water Rescue Training Manual-Equipment)

    6.    Make proper water entry, for minimal turbidity and maximum safety.

    7.    Search Datum Point upon initial entry.

    8.    Establish line length, based on visibility. Line loop shall be hand held for quick release (see Water Rescue Training Manual).

    9.    Begin surface directed search.

Note:        Surface Directed Search is the safest for diver accountability and is the most efficient.

B.    Safety Diver (wearing mandatory and accessory equipment (see Attachment #1)

       1.    Stand by at water's edge

       2.    Render assistance to Rescue Skin Diver as necessary

C.    Line Tender (with BC & foot protection) controls surface directed search.

       1.    Use appropriate signals to inform diver:

            a.    Reaching limit of Area of Probability
            b.    Termination of search

       2.    Respond to Rescue Skin Diver's line signals.

       3.    Maintain integrity of Search Pattern by maintaining line length, unless signaled for more line or pull in.

       4.    Maintain safety of Rescue Skin Diver by awareness of diver location and response to line signals.

       5.    Inform Safety Diver if Primary Diver needs assistance.

       6.    Monitor hand-held radio.

# III. Search, Extrication and Rescue of a Drowning Victim, from a vehicle in the water.

On all night operations, all personal floatation (BC or PFD) shall have a cyalume light stick attached and activated.

A.    Rescue Skin Diver (with mandatory and accessory equipment see Attachment #1, including center punch and seat belt cutter)

    1.    Determine Datum Point.
    2.    Establish Area of Probability.
    3.    Confirm type of Search Pattern .

Note:    If a known Datum Point is available a Circular Pattern should be implemented. The use of a Linear Pattern should be employed when a Datum Point is not known.

    4.    Confirm dive plan & line signals with Line Tender OATHS

            One Pull =        OK
    a.    Two Pulls =      Advance Line (Give Me Line)
            Three Pulls =    Take up Line (Pull Me In)
    d.    Four Pulls =     Help! (Secure Line & Assist Me) (1)
    e.    Five Pulls =

        ( 1)    If help signal is received, tie loop in line to mark diver distance, note direction to diver; alert safety diver to respond.

    5.    Rig torpedo buoy on water rescue line (see Water Rescue Training Manual).

    6.    Make proper water entry for minimal turbidity.

    7.    Search Datum Point upon initial entry.

    8.    Establish line length, based on visibility. Line loop shall be hand held for quick release (see Water Rescue Training Manual).

    9.    Begin Surface Directed Search .

Note:    Surface Directed Search is the safest for diver accountability and is the most efficient.

    10.    When location of vehicle is established, the marker float shall be deployed for permanent reference.

11.   Establish cautious entry into vehicle if necessary.

Note:       Use of center Punch for forcible entry may be employed. The
Rescue Diver shall:

   a.   Wear Gloves
   b.   Locate comer of window (not windshield)
   c.   Brace hand on window frame
   d.   Repeatedly 'pop' glass in the same spot until glass
        breaks
   e.   Clear glass from frame with gloved hand and/or tool

Note:       If Unconscious victim is located in the vehicle, free victim
from seat belt or entanglement. The patient should be extricated by the
Long Axis of the body, (head, arm, both legs). The patient should be Head
Carried to surface and then converted to Cross Chest Carry to shore. (Use
of B.C. should be employed as necessary).

Note:       If Conscious Victim is located in the vehicle, let the patient
know you are there, avoid physical contact. If the patient exits vehicle on
his own, use Heimlich Carry to surface patient, and employ the Cross
Chest Carry to shore. (Use of B.C. should be employed as necessary).

Note:       If the patient remains in trapped air pocket, and it is found
necessary to extricate patient by other means, enrich air pocket with
SCBA tank. Make repeated dives on vehicle, allowing sufficient surface
recovery time between dives. If patient becomes unconscious revert to
Unconscious Victim Procedure. After recovery of a viable drowning
patient, conduct appropriate in-water medical treatment as indicated.
Remove patient from water with assistance of shore personnel. Use
adjuncts when possible (roof ladder aerial units, etc.). Treat patient per
appropriate protocol.

B.   Safety Diver wearing mandatory and accessory equipment (see
     Attachment #1).

   1.   Stand by at water's edge.
   2.   Render assistance to Rescue Skin Diver as necessary.

C.   Line Tender (with BC & foot protection) controls surface directed
     search.

   1.   Use appropriate signals to inform diver:

      a.   Reaching limit of Area of Probability
      b.   Termination of search

2.    Respond to Rescue Skin Diver's line signals.

3.    Maintain integrity of Search Pattern by maintaining line length, unless signaled for more line or pull in.

4.    Maintain safety of Rescue Skin Diver by awareness of diver location and response to line signals.

5.    Inform Safety Diver if Primary Diver needs assistance.

6.    Monitor hand-held radio.

## IV. Offshore search for a drowning victim or vehicle (non-surface directed, two man)

Note:       On all night operations, all personal flotation (BC or PFD) shall have a cyalume light stick attached and activated.

A.       Two Man Rescue Skin Diver Buddy Team with Mandatory and accessory equipment (see Attachment # 1) including center punch & seat belt cutter; additional marker float recommended.

    1.       Determine Datum Point.
    2.       Establish Area of Probability.
    3.       Confirm type of Search Pattern.

Note:       If a known Datum Point is available a Circular Pattern should be implemented.

    4.       Confirm Dive Plan.

    5.       Make proper water entry, for minimal turbidity and maximum safety.

    6.       Deploy Marker buoy, at Datum Point (Center of area of probability).

    7.       Breath-hold dive Datum Point first. Alternate Divers: one up; one down for safety reasons.

    8.       Expand search outward from Datum Point in concentric circular dives. Alternate Divers: One diver up; one diver down for safety reasons.

    9.       When location of vehicle is established the marker float shall be deployed for permanent reference.

    10.       Establish cautious entry into vehicle, if necessary.

Note:       Use of center Punch for forcible entry may be employed. The Rescue Diver shall:

        a.       Wear Gloves
        b.       Locate comer of window (not windshield)
        c.       Brace hand on window frame
        d.       Repeatedly 'pop' glass in the same spot until glass breaks
        e.       Clear glass from frame with gloved hand and/or tool

Note:      If Unconscious victim is located in the vehicle, free victim from seat belt or entanglement. The patient should be extricated by the Long Axis of the body, (head, arm, both legs). The patient ,should be Head Carried to surface and then converted to Cross Chest Carry to shore. (Use of B.C. should be employed as necessary).

Note:      If Conscious Victim is located in the vehicle, let the patient know you are there, avoid physical contact. If the patient exits vehicle on his own, use Heimlich Carry to surface patient, and employ the Cross Chest Carry to shore. (Use of B.C. should be employed as necessary),

Note:      If the patient remains in trapped air pocket, and it is found necessary to extricate patient by other means, enrich air pocket with SCBA tank. Make repeated dives on vehicle, allowing sufficient surface recovery time between dives. If patient becomes unconscious revert to Unconscious Victim Procedure. After recovery of a viable drowning patient, conduct appropriate in-water medical treatment as indicated. Remove patient from water with assistance of shore personnel. Use adjuncts when possible (roof ladder aerial. units, etc.). Treat patient per appropriate protocol.

# WATER RESCUE OPERATING PROCEDURES FOR SCUBA RESCUE RESPONSE

I.     Search and Rescue of a Drowning Victim -Surface Directed From Shore

(SCUBA Rescue Operations should not be undertaken unless a minimum of two department SCUBA Authorized divers are present). On all night operations, all personal flotation (BC or PFD) shall have a cyalume light stick attached and activated.

A.     Department Authorized Scuba Rescue Diver (wearing mandatory and accessory equipment; see Attachment #1).

      1.     Determine Datum Point
      2.     Establish Area of Probability
      3.     Confirm type of Search Pattern

Note:     If a known Datum Point is available a Circular Pattern should be implemented. The use of a Linear Pattern should be employed when a Datum Point is not known.

      4.     Confirms Rescue Line signals with Line Tender: OATHS

          a.     One Pull =          OK
          b.     Two Pulls =        Advance Line (Give Me Line)
          c.     Three Pulls =     Take up Line (Pull Me In)
          d.     Four Pulls =       Help! (Secure Line & Assist Me) (1)
          e.     Five Pulls =        Stop Search/surface

            ( 1)   If help signal is received, tie loop in line to mark diver distance, note direction to diver; alert safety diver to respond.

      5.     Rig torpedo buoy on water rescue line, (see Water Rescue Training Manual-Equipment).

      6.     Make proper water entry, for minimal turbidity and maximum safety.

      7.     Search Datum Point, upon initial entry.

      8.     Establish line length, based on visibility. Line loop, shall be hand held for quick release (see Water Rescue Training Manual).

      9.     Begin Surface Directed Search.

Note: Surface directed search is the safest for diver accountability and is the most efficient.

B. Safety Diver (wearing mandatory and accessory equipment, see Attachments #1).

Note: Safety Diver shall be a Department Scuba Authorized Diver. He may assist the SCUBA Authorized Diver as a Stand-by Safety Diver, on shore, with a second set of Scuba gear, geared up and ready to enter the water.

C. Line Tender (with BC and foot protection - controls Surface Directed Search).

1. Use appropriate signals to inform diver.

   Reaching limit of Area of Probability
   6: Termination of search

2. Respond to Department SCUBA Authorized Diver's line signals.

3. Maintain integrity of Search Pattern by maintaining line length, unless signaled for more line or pull in.

4. Maintain safety of Department SCUBA Authorized Diver by awareness of diver location and response to line signals.

5. Inform Safety Diver if Primary SCUBA Diver needs assistance.

6. Monitor hand-held radio.

7. Check status of Authorized SCUBA Rescue Diver periodically with line signals.

II.   Search, Extrication and Rescue, of a drowning victim, from a vehicle in the water. Surface Directed From Shore.

Note:      On all night operations, all personal flotation (BC or PFD) shall have a cyalume light stick attached and activated.

A.   Department SCUBA Authorized Diver (wearing mandatory and accessory equipment, see Attachment #1, including center punch and seat belt cutter).

   1.   Determine Datum Point.
   2.   Establish Area of Probability.
   3.   Confirm type of Search Pattern.

Note:      If a known Datum Point is available a Circular Pattern should be implemented. The use of a Linear Pattern should be employed when a Datum Point is not known.

   4.   Confirm Dive Plan and Line signals with Line Tender: OATHS

      a.   One Pull =      OK
      b.   Two Pulls =     Advance Line (Give Me Line)
      c.   Three Pulls =   Take up Line (Pull Me In)
      d.   Four Pulls =    Help (Secure Line and Assist Me) (1)
      e.   Five Pulls =    Surface

      (1)   If help signal is received, tie loop in line to mark diver distance, note direction to diver; alert safety diver to respond.

   5.   Rig torpedo buoy on water rescue line (see Water Rescue Training Manual-Equipment).

   6.   Make proper water entry for minimal turbidity and maximum safety.

   7.   Search Datum Point upon initial entry.

   8.   Establish line length, based on visibility (line loop, hand held for quick release see Water Rescue Training Manual).

   9.   Begin Surface Directed Search.

Note:      Surface Directed Search is the safest for diver accountability and is the most efficient.

THIS PAGE LEFT INTENTIONALLY BLANK

10. When location of vehicle is established the marker float shall be deployed for permanent reference.

11. Establish cautious entry into vehicle, if necessary.

Note: Use of Center Punch for forcible entry may be employed. The Rescue Diver shall:

**a)** Wear Gloves
b) Locate corner of window (not windshield)
c) Brace hand on window frame
d) Repeatedly 'pop' glass in the same spot until glass breaks
e) Clear glass from frame with gloved hand and/or tool.

Note: If Unconscious Victim is located in the vehicle, free victim from seat belt or entanglement. The Patient should be extricated by the Long Axis of the body (head, arm, both legs). The patient should be Head Carried to surface and then converted to Cross Chest Carry to shore (B.C. should be employed as necessary).

Note: If Conscious Victim is located in the vehicle, let the patient know you are there, avoid physical contact. If the patient exits vehicle on his own, use Heimlich Carry to surface patient, employ the Cross Chest Carry to shore (B.C. should be employed as necessary).

Note: If the patient remains trapped in air pocket and it is found necessary to extricate patient by other means, enrich air pocket with SCBA tank. If patient becomes unconscious revert to Unconscious Victim Procedure. After recovery of a viable drowning patient, conduct appropriate in-water medical treatment as indicated. Remove patient from water with assistance of shore personnel. Use adjuncts when possible (roof ladder, aerials, etc.). Treat patient per appropriate protocol.

# III. Quayside search for a drowning victim or vehicle, surface directed from dock or quay.

Note: On all night operations, all personal flotation (BC or PFD) shall have a cyalume light stick attached and activated.

A. Department SCUBA Authorized Diver (with mandatory and accessory equipment (see Attachment #1), including center punch and seat belt cutter.

1. Determine Datum Point.
2. Establish Area of Probability,
3. Confirm type of Search Pattern.

Note: If a known Datum Point is available a Circular Pattern should be implemented.

4. Confirm Dive Plan and Line Signals: OATHS

a. One Pull =          OK
b. Two Pulls =         Advance Line (Give Me Line)
c. Three Pulls =       Take up Line (Pull Me In)
d. Four Pulls =        Help (Secure Line and Assist Me) (1)
e. Five Pulls =        Surface

(1) If help signal is received, tie loop in line to mark diver distance, note direction to diver; alert safety diver to respond.

5. Lower *anchor" to bottom at datum point and secure "anchor line". Tender stands at this anchor point above anchor on quay or dock.

6. Make proper water entry for minimal turbidity and maximum safety.

7. Search Datum Point, at anchor, upon initial entry.

8. Run search line through carabiner or large snap attached to anchor. Using anchor as a pivot point, establish line length based on visibility (line loop hand held for quick release, see Water Rescue Training Manual).

9. Begin Surface Directed Circular (Arc) Search.

Note: Surface Directed Search is the safest for diver accountability and is the most efficient.

3:19

10. When location of vehicle: is established the marker float shall be deployed for permanent reference.

11, Establish cautious entry into vehicle, if necessary.

Note: Use of Center Punch for forcible entry may be employed. The Rescue Diver shall:

    a) Wear Gloves
    b) Locate comer of window (not windshield)
    c) Brace hand on window frame
    d) Repeatedly 'pop' glass in the same spot until glass breaks
    e) Clear glass from frame with gloved hand and/or tool.

Note: If Unconscious Victim is located in the vehicle, free victim from seat belt or entanglement. The Patient should be extricated by the Long Axis of the body (head, arm, both legs). The patient should be Head Carried to surface and then converted to a Cross Chest Carry to shore (B.C. should be employed as necessary).

Note: If Conscious Victim is located in the vehicle, let the patient know you are there, avoid physical contact. If the patient exits vehicle on his own, use Heimlich Carry to surface patient, employ Cross Chest Carry to shore (B.C. should be employed as necessary).

Note: If the patient remains trapped in air pocket and it is found necessary to extricate patient by other means, enrich the air pocket with SCBA tank. If patient becomes unconscious revert to Unconscious Victim Procedure. After recovery of a viable drowning patient, conduct appropriate in-water medical treatment as indicated. Remove patient from water with assistance of shore personnel. Use adjuncts when possible (roof ladder, aerials, etc.). Treat patient per appropriate protocol.

## IV.	Offshore search for a drowning victim or vehicle, surface directed from a boat.

Note:	On all night operations, all personal flotation (BC or PFD) shall have a cyalume light stick attached and activated.

A.	Department Authorized Scuba Rescue Diver (with mandatory and accessory equipment (see Attachment #1) including center punch and seat belt cutter).

1.	Determine Datum Point
2.	Establish Area of Probability
3.	Confirm type of Search Pattern

Note:	If a known Datum Point is available a Circular Pattern should be implemented.

In calm water, boat should be anchored at the Datum Point. In a current situation, boat should be anchored up current of the Datum Point and a cross current "Fan Sweep" can be made astern starting at the down current limit of the Area of Probability and working toward the boat. (see Water Rescue Training Manual)

4.	Confirm Rescue Line signals with Line Tender: OATHS

a.	One Pull =	OK
b.	Two Pulls =	Advance Line (Give Me Line)
c.	Three Pulls =	Take up Line (Pull Me In)
d.	Four Pulls =	Help! (Secure Line & Assist Me)
e.	Five Pulls =

Note:	On all night operations, all personal flotation (BC or PFD) shall have a cyalume light stick attached and activated.

5.	Make proper water entry, for minimal turbidity and maximum safety.

6.	Search Datum Point, upon initial entry.

7,	Establish line length, based on visibility. Line loop, shall be hand held for quick release (see Water Rescue Training Manual).

8.	Run search line through anchor ring. Anchor will be pivot point for search pattern.

9.	Begin Surface Directed Search.

10.    When using a full circle pattern, reverse direction with each successive sweep to avoid line entanglement.

Note:    Surface directed search is the safest for diver accountability and is the most efficient.

B.    Safety Diver (wearing mandatory and accessory equipment, see Attachment #1).

Note:    Safety Diver shall be a Department Scuba Authorized Diver, he may assist the Authorized SCUBA Diver as a Stand-by Safety Diver, on the boat with a second set of Scuba gear, geared up and ready to enter the water.

C.    Line Tender (with BC or PFD ) controls Surface Directed Search.

1.    Use appropriate signals to inform diver.

a.    Reaching limit of Area of Probability
b.    Termination of search

2.    Respond to Department SCUBA Authorized Diver's line signals.

3.    Maintain integrity of Search Pattern by maintaining line length, unless signaled for more line or pull in.

4.    Maintain safety of Department SCUBA Authorized Diver by awareness of diver location and response to line signals.

5.    Inform Safety Diver if Primary SCUBA Diver needs assistance.

6.    Monitor hand-held radio.

7.    Check status of Authorized SCUBA Rescue Diver periodically with line signals.

v.   Offshore search for a drowning victim or vehicle. (none surface directed, two-man diver directed search).

Note:         On all night operations, all personal flotation (BC or PFD) shall have a cyalume light stick attached and activated.

A.   DEPARTMENT SCUBA AUTHORIZED DIVERS ( with Mandatory and accessory equipment (see Attachment # 1, including center punch and seat belt cutter); additional marker float recommended.

  1.   Determine Datum Point.
  2.   Establish Area of Probability.
  3.   Confirm type of Search Pattern.

Note: (OPTION 1) - From a Datum Point the dive team may cover the area of probability using a Circular Pattern employing one Dive Team member as a Datum Point Anchor / Tender and the other Dive Team member as a Primary Search Diver ( See Module 12 of the Water Rescue Training  Manual)

  4.   Confirms Dive Plan and Line Signals: OATHS

      a.   One Pull =        OK
      b.   Two Pulls =       Advance Line (Give Me Line)
      C.   Three Pulls =     Take up Line (Pull Me In)
      d.   Four Pulls =      Help! (Secure Line & Assist Me)
      e.   Five Pulls =      Surface

NOTE:      On all night operations, all personal flotation (BC or PFD) shall have a cyalume light stick attached and activated.

  5.   Make proper water entry, for minimal turbidity and maximum  safety.
  6.   Deploy marker float at Datum Point.

  7.   Submerge at Datum Point and check bottom for victim.

  8.   Establish line length, based on visibility.

      a.   Line Container, hand held by Primary Search Diver for convenience.

      b.   Line loop shall be hand held by Anchor/Tender (see Water Rescue Training Manual).

  9.   Begin Diver Directed Search.

10. When location of vehicle is established a marker float shall be deployed for permanent reference.

11. Primary Search Diver Signals Anchor/Tender to join him (Three Pulls).

12. Establish cautious entry into vehicle, if necessary.

Note: Use of Center Punch for forcible entry may be employed. The Rescue Diver shall:

   a) Wear Gloves
   b) Locate comer of window (not windshield)
   c) Brace hand on window frame
   d) Repeatedly 'pop' glass in the same spot until glass breaks
   e) Clear glass from frame with gloved hand and/or tool

Note: If Unconscious Victim is located in the vehicle, free victim from seat belt or entanglement. The Patient should be extricated by the Long Axis of the body (head, arm, both legs). The patient should be Head Carried to surface and then converted to Cross Chest Carry to shore (B.C. should be employed as necessary).

Note: If Conscious Victim is located in the vehicle, let the patient know you are there, avoid physical contact. If the patient exits vehicle on his own, use Heimlich Carry to surface patient, employ the Cross Chest Carry to shore (B.C. should be employed as necessary).

Note: If the patient remains trapped in air pocket and it is found necessary to extricate patient by other means, enrich air pocket with SCBA tank. If patient becomes unconscious revert to Unconscious Victim Procedure. After recovery of a viable drowning patient, conduct appropriate in-water medical treatment as indicated. Remove patient from water with assistance of shore personnel. Use adjuncts when possible (roof ladder, aerial units, etc.). Treat patient per appropriate protocol.

Note: (OPTION 2) - From a Datum Point the dive team may cover the Area of Probability using a Compass Guided Expanding Square Search. A compass will be required as accessory equipment (see Water Rescue Training Manual).

# VI. Water Rescue operating procedures for rescue diver air delivery

A.  Deployment of Rescue Divers to an Offshore incident will involve only department SCUBA Authorized Divers who have completed the "Diver Air Delivery P.I.T. Class".

    1.  Determine Dive Team Leader.
    2.  Determine Drop Point.
    3.  Mandatory and Accessory Equipment; (See Attachment # 1 including center punch and seat belt cutter).

Note:    On all night operations, all personal flotation (BC or PFD) shall have a cyalume light stick attached and activated.

    4.  Minimum 2 Man-Team Deployed, at discretion of Dive Team Leader; Buddy Contact Maintained.

    5.  Divers deployed in Rescue Skin Diver or SCUBA Rescue Diver mode, as appropriate.

    6.  Team Leader (First to Deploy) makes the Dive/No-Dive Decision.

B.  Deployment procedures must be in accordance with training presented in " Diver Air Delivery " P.I.T. Class, and Transporting agency S.O.P.'s ( Air Rescue Division, U.S.C.G., Ect...)

## VII.  Personal flotation device use (PFD):

A.  All personnel working in close proximity to the water or on board water craft will wear a PFD if available or a BC as supplied on FD units.

B.  BCs should be adjusted for fit and support and must have an operable $CO_2$ cartridge for rapid inflation. They must be equipped with a whistle.

C.  Additional BCs and PFDs can be obtained on a large water rescue incident or near-water fire scene by responding Water Rescue Bureau staff units for support.

Note:    On all night operations, all personal flotation (BC or PFD) shall have a cyalume light stick attached and activated.

# VIII.    Diver status

A.    SCUBA Rescue Authorized Divers shall be assigned an "activity status" based on their level of participation in the Water Rescue Program.

    1.    Active Diver - SCUBA Rescue Divers who have:

        a.    Attended 3 PIT classes per year.
        b.    Submitted a Dive Log reflecting 12 dives/year.

    2.    Inactive Diver - SCUBA Rescue Divers who are interested in continued involvement in the Water Rescue Program but who are unable to qualify as "Active Diver" as described above.

    3.    Terminated - Personnel who are no longer interested in participating in the Water Rescue Program or who have been removed from "Active- or Inactive Diver" status for health or safety reasons.

B.    Only divers on Active Diver status can qualify for advanced and specialty diver certifications or other benefits or incentives of the Water Rescue Program.

C.    All Active Divers shall be issued a SCUBA Rescue Authorized Diver Certification card, updated annually.

## IX. Dive logs:

A. Are due annually, by January 31, for the preceding year.

B. Logs must reflect a minimum of 12 dives/year on SCUBA. Six of these dives must be Search and Rescue related.

C. Dive logs will be turned in to the Water Rescue Bureau for evaluation, computer entry and diver file entry.

D. Dive Log Sheets are available from the Water Rescue Bureau Diving Office or can be found in the Water Rescue Training Manual.

## x. Water incident report:

A.    This report form is to be filled out following a water incident, by any fire department unit that deploys divers. Water entry, regardless of rescue results, is the primary factor governing  use of this form.

B.    Water Incident Reports are available in the stations, in the Water Rescue Training Manual, or through Supply.

## XI.   Touch signals:

A.   In conditions of limited visibility, underwater communication by members of a two-man dive team shall be accomplished by touch signals.

B.   Signals will be by arm squeeze.

C.   Touch signals will replicate water rescue line signals:

1.   One Squeeze =      OK
2.   Two Squeezes =     Move Away from me
3.   Three Squeezes =   Come to (with) me
4.   Four Squeezes =    Help!
5.   Five Pulls =       Surface

D.   Out Of Air Signal: Four Squeezes and tap buddy's SCUBA regulator second stage.

(ATTACHMENT #1)

WATER RESCUE EQUIPMENT LIST

1.  Mandatory Rescue Skin Diver Equipment
    a.  Mask
    b.  Fins
    c.  Snorkel
    d.  B.C. with cyalume light stick & whistle (for line tender too)
    e.  Water Rescue Line and/or Torpedo Buoy

2.  Mandatory Scuba Rescue Equipment
    a.  Mask
    b.  Fins
    c.  Snorkel
    d.  B.C. with cyalume light stick & whistle (for line tender too)
    e.  Scuba Tank (minimum 3/4 full of air or 2250 psi)
    f.  Regulator with S.P.G. (submersible pressure gauge)
    g.  Water Rescue Line (with torpedo buoy attached for surface directed search)

3.  Personal Protection Equipment
    a.  Uniform
        1)  Duty Uniform
        2)  Dive Skin
        3)  Wet Suit
        4)  Dry Suit
    b.  Protective Gear
        1)  Foot Gear (shoes, booties)
        2)  Gloves

4.  Accessory Rescue Equipment: Gear used to aid in rescue effort.
    a.  Underwater Search Light
    b.  Automatic Center Punch
    c.  Boot Shears or knife
    d.  Compass
    e.  Marker Float
    f.  Rescue Strobe
    g.  Dive Tables-current U.S. Navy model, conservatively used.
    h.  Dive Timer-digital or analog
    i.  Depth gauge-oil filled
    j.  Dive Computer-used as a back-up to dive tables (see Water Rescue Training Manual)

Accessory Equipment which is not department issue must be approved by the Water Rescue Bureau and will be logged in "Personal Equipment Inventory" form in Diver File.

(ATTACHMENT #2)

INSPECTION AND CARE OF SCUBA EQUIPMENT

A.  Morning Equipment Check (Pre-Dive Check) by Authorized SCUBA Rescue Diver;

    1) Backpack and Harness: no defects and secure

    2) Regulator and Hoses: no defects and secure

    3) Air on: no leaks
       a) Cheek PSI on S.P.B. (tank should be 3/4 full or above
       b) Check Breathe Regulator (3 hard breaths)
       c) Taste air for quality

    4) Check mandatory equipment: mask, fins, snorkel, B.C.; Water Rescue Line.

    5) Check accessory equipment: Gloves, Light, Automatic Center Punch, Knife, etc.

    6) Check personal protection equipment and as appropriate.

B.  Following return to quarters:

    1) Clean all Water Rescue equipment with fresh water

    2) Air-dry equipment before storage

    3) Drain B.C., install new $CO_2$ if necessary, partially inflate and secure on unit

    4) Dry and lubricate Center Punch

    5) Unlock Mask and Fin Straps before storage

    6) Divers should blow dry ears to prevent possibility of Swimmers Ear irritation

    7) Pre-dive check SCUBA equipment as in Morning Equipment Check, this protocol. (SCUBA tank must be 3/4 full or above to be in service: 2250 psi)

Note:  Any Changes of location of equipment must be approved in writing by the Water Rescue Bureau and the Safety Office.

C.   Consult Station Water Rescue Training Manual.
For Guidelines on:
1.   Scuba Air Fill Procedures
2.   Equipment Repair & Replacement
3.   SCUBA Cylinder Visual Inspection & Hydrostatic    Tests.
4.   Equipment Maintenance
5.   Location of equipment stored on units
6.   Rigging procedures for torpedo buoy in Surface Directed Searches
7.   Water Rescue Line hand-hold technique
8.   Special Equipment and capabilities of Water Rescue Bureau staff units
9.   Diving Accident Management procedure
10.   Rubber Inflatable Boat (RIB) SOP
11.   R.A.F.T. (Hose Boom) Deployment
12.   BC rigging
13.   Mixed Gas Diving (NITROX) SOP
14.   Surface Air Supply Diving SOP
15.   Dry Suit Diving SOP
16.   Water Rescue Training Boat (Hard Hull) S.OP

(ATTACHMENT #3)

## GLOSSARY OF CERTIFICATION LEVELS

1.  Rescue Skin Diver Certified: Certified by NAUI and Metro-Dade Fire Department as a Rescue Skin Diver.

2.  SCUBA Certified: Certified by a National Diver Training Agency (NAUI,-PADI, YMCA, NASDS, SSI, L. A. County etc) as having passed a SCUBA Training Course, Basic Level or higher.

3.  SCUBA Rescue Authorized: Certified by Metro-Dade Fire Department D.R.T.T. to use SCUBA on duty.

4.  Rescue SCUBA Diver Certified: Certified by NAUI and Metro-Dade Fire Department. this specialty SCUBA certification requires department SCUBA Authorization as a prerequisite and additional specialty Dive Rescue Training.

5.  SCUBA Search and Recovery Certification: Certified by NAUI, PADI, YMCA, NASDS, SSI, LA County, etc. and Metro-Dade Fire Department. This advanced specialty certification requires Rescue SCUBA Diver Certification as a prerequisite and additional specialty training.

6.  Skin Diving Leader Certification: Certified by NAUI. This instructor level certification requires SCUBA Search and Recovery Certification as a prerequisite. It also requires additional specialty training and teaching experience in water rescue.

7.  Divemaster Certification: Certified by NAUI, PADI, YMCA, etc.. This advanced certification requires Rescue SCUBA Diver Certification as a prerequisite. It also requires additional specialty training and experience in Divemaster techniques

8.  Assistant Instructor Certification: Certification by NAUI, PADI, YMCA, etc.. This instructor level certification requires SCUBA Search and Recovery Certification as a prerequisite as well as additional specialty training and teaching experience.

9.  SCUBA Instructor Certification: Certified by NAUI, PADI, YMCA, etc. This Instructor certification requires Assistant Instructor Certification as a prerequisite, and requires successful completion of an Instructor ITC (Instructor Training Course).

10. SCUBA Search & Recovery Instructor Certification: Certified as a SCUBA S&R Instructor by NAUI, PADI, YMCA, etc.

11. Rescue Boat Operator Certification: Certified by D.R.T.T. or Marine Rescue Consultants for Rescue Training Boat Operation.

12. U.S.Coast Guard Licensed Captain: Certified by USCG, six passenger or above, for boat operation.

13. American Red Cross Swimming and Lifesaving Certifications: ARC Water Safety Instructor, Lifeguard Instructor and Beginner Swimmer through Lifeguard Certified.

14. NITROX Certified: Certified by Hyperbarics International, Inc. trained & certified instructor to dive NOAA Nitrox I and Nitrox II.

(ATTACHMENT #4)

## GLOSSARY OF WATER RESCUE TERMS

1.    Area of Probability: Limited area where a lost object is most likely to be found.

2.    BC: Buoyancy Compensator, a variable volume item of Mandatory Equipment for self-rescue, patient rescue and buoyancy control.

3.    Boat Directed Search: Offshore search pattern surface directed by tender in boat with safety diver equipped & standing by on board.

4.    Carries: Means of moving a drowning victim, under control of the rescuer, to a place of safety.

5.    Cyalume: Chemical light stick, attached to BC neck by 18 inch long lanyard, shall be activated and displayed on all night operations (see Water Rescue Training Manual).

6.    Center Punch: Automatic Center Punch for breaking side and rear windows of a vehicle.

7.    Circular Pattern: Circular or semi-circular coverage of Area of Probability when 'Datum Point' is known.

8.    'Datum Point': Last known position of lost object.

9.    D.R.T.: Dive Rescue Team refers to SCUBA Rescue Authorized Divers on FD units.

10.   D.R.T.T.: Dive Rescue Training Team refers to volunteer part-time Dive Rescue Instructors training D.R.T. Divers through Water Rescue Bureau.

11.   Expanding Square Search: Compass directed search pattern for offshore use (see Water Rescue Training Manual, Module 12).

12.   Fan Sweep: A Semi-Circular variation of a Circular Search Pattern, shore based or boat based, surface directed for down-current search use.

13.   Linear Pattern: Straight line pattern for coverage of an Area of Probability when Datum Point is not known.

14. Line Signals: Communications by diver and tender on Search Line by means of pulls.

15. Line Rescue: Water rescue evolution using Water Rescue Line, Rescue Skin Diver and Line Tender.

16. Line Tender: Directs search from shore via Search Line or pulls rescuer and victim to shore in surface rescue by means of Water Rescue Line.

17. Marine Rescue Consultants: A provider of Rescue Boat Training, 1825 Westcliff Drive, Suite #105, Newport Beach, Ca. 92660

18. N.A.U.1: National Association of Underwater Instructors, an international diver training agency.

19. NITROX: Oxygen enriched gas mixture as defined by NOAA standards for increased physiological or time advantage in dive operations deeper than 60 fsw.

20. P.A.D.I.: Professional Association of Diving Instructors, an international diver training agency.

21. P.I.T.: Proficiency Improvement Training classes provided monthly, on and off-duty for SCUBA Rescue Authorized Diver training.

22. PFD: Personal Flotation Device, USCG Approved, Class III.

23. Quayside Search: A semi-circular, surface directed search pattern conducted alongside a quay or dock.

24. Safety. Diver: A Back-Up diver, geared up, on shore, available to assist.

25. SCUBA: Self Contained Underwater Breathing Apparatus.

26. Search Line: Water Rescue Line, 100' three strand twisted 1/4 inch polypropylene with a monkey's fist at one end and 30" loop with snap at the other end.

27. Search Pattern: Logical means of covering a Search Area while maintaining diver accountability (line tender or buddy).

28. SSI: SCUBA Schools International, a national diver training agency.

29. Torpedo Buoy: Towable Surface Rescue Float (Can Buoy, Torp, Peterson Buoy, Rescue Can, etc.)

30. Torpedo Buoy Rescue: Water Rescue evolution using a Torpedo Buoy and a Rescue Skin Diver.

31. Unassisted Rescue: Water rescue evolution conducted by a Rescue Skin Diver without assistance of a Torpedo Buoy or water Rescue Line.

32. Water Rescue Evolution: A means of conducting a rescue of an actively drowning victim on the surface.

33. Water Rescue Line: See Search Line

34. Water Sector: That part of a water rescue scene that involves department divers and dive rescue activity. It is the area of responsibility of the Senior Rescue Diver, working within Incident Command Procedures.

35. Y.M.C.A.: Young Men's Christian Association, a national diver training agency.

THINGS THAT STILL NEED TO BE WRITTEN ABOUT
Rubber Inflatable Boat (RIB) SOP
Water Rescue Training Boat SOP
Mass casualty situations
Raft deployment
Contaminated water testing
Lake surveys
Rescues off piers and bridges
Handling cervical injuries in water
Surf/beach/current rescues
Swift water rescues
Marine firefighting capabilities
Other diving situation training
DRTT instructor command capabilities

26.06 REVOCATION:
A.O. 41-89 and all parts of previous orders, rules and regulations, operations memos and administrative orders in conflict with this policy and procedure.

# METROPOLITAN DADE COUNTY FIRE RESCUE DEPARTMENT
## WATER RESCUE BUREAU

### NITROX STANDARD OPERATING PROCEDURES (SOP)

**Introduction:** This SOP is established to provide guidelines for proper practice of "Enriched Air" diving. The "Enriched Air" gases in question are mixtures of Oxygen and Nitrogen, with resulting percentages ranging from 30% to 50% Oxygen and the balance Nitrogen, otherwise referred to as Nitrox.

**Purpose:** To provide divers a safer breathing medium for dives in the 40' to 130' depth range for the purpose of training or operational use. The Fire Department's Dive Officer may elect to place divers on Nitrox for either the physiological advantage or the decompression advantage, see footnote #I, depending on the nature of the dive.

**Mixtures:** Nitrox gas mixes were developed by the National Oceanographic and Atmospheric Administration, (NOAA) and are endorsed by the International Association of Nitrox Divers (IAND), the American Nitrox Divers, Inc. (ANDI), National Association of SCUBA Diving Schools (NASDS) and National Association of Underwater Instructors (NAUI). The two standard mixtures to be utilized by authorized divers shall be NOAA Nitrox I, (NNI), (see footnote #2), and NOAA Nitrox II, (NNII), (see footnote #3).

NNI is most useful for dives in the 100 to 130 range and NNII should be used diving the 40' to 100' range. When deemed necessary, other mixtures may be utilized under the direction and at the discretion of the Department Dive Officer. Furthermore, all guidelines and standards established by NOAA, IAND, and AND1 will be strictly adhered to when "brewing" other mixtures.

**Authorized Divers:** A Nitrox diver list is kept on file at the Fire Department's Water. Rescue Bureau indicating those divers qualified and authorized for Nitrox use. Any SCUBA Rescue Authorized (SRA) diver requesting inclusion on that list, will be required to show proof of Nitrox Diver Certification by Hyperbarics Int'l., IAND or ANDI and/or pass a written exam administered by the Water Rescue Bureau pertaining to Nitrox diving. The use of Nitrox is restricted to only those divers on the Nitrox list.

**Nitrox Cylinders:** Only cylinders dedicated exclusively to the use of Nitrox and properly marked according to Dive Rescue Training Team, (DRTT) guidelines, will be allowed for Nitrox Use, (see figure #1).

**Fills:**    All Nitrox cylinders must be filled by facilities approved by the Fire Department's Water Rescue Bureau. All fills obtained at the Department's Dive Locker will be done by either by the Dive Officer, Capt. Ed Brown or the Assistant Dive Officer, Herb Smith, with strict adherence to established fill procedures. As soon as filling is completed, each cylinder will be tested for Oxygen level using an Oxygen analyzer and marked to indicate the mix and % of Oxygen.

**Nitrox Fill Log:**    An active cylinder log is kept at the Dive Locker Fill Station to record all Nitrox fills. Information to be entered into the log immediately after each fill is as follows;

1.  Date
2.  Cylinder Serial Number
3.  Volume
4.  Mix
5.  Percentage of Oxygen, (see footnote #4)
6.  Total PSI
7.  Verification
8.  Date of Use
9.  Location of fill (if not at Dive Locker)
10. Name and Signature of Filler

If the cylinder was filled at a facility other than the Dive Locker the above information will be entered as soon as possible, including the location of fill. Before any cylinder can be used, the individual diver or the person responsible for transporting the cylinder to the dive site, must initial the fill log verifying that the information entered is accurate. A second analysis of the mix may be necessary.

Each diver is responsible for testing the mix in their cylinder immediately prior to the dive to ensure proper compliance with NOAA regulations, (see footnote #4). The diver will enter and initial the test results in the Divemaster's Log, (see inclusion #1).

To ensure proper quality assurance of all Nitrox dives, Divemaster's Logs must be submitted to the Water Rescue Bureau within 72 hours of completion of dive operations.

Footnotes:

1.  NOAA Nitrox tables will be utilized by the Divemaster.
2.  32% Oxygen/68% Nitrogen
3.  36% Oxygen/64% Nitrogen
4.  NOAA regulations require that the Oxygen percentages be within +/- 1% of recommended levels to be acceptable.

# WATER RESCUE RECORD

| ID SECTION |
|---|

**Record:**            This identifies the record for the computer.

**Unit #:**            identify the unit handling the incident. See Table 6.

**Shift:**            Enter A, B, or C for the shift of the unit handling the call.

**Department:**            Two digit department code of your department. See Table 1.

**Incident #:**            Also called Alarm number. Sequential number assigned to each incident by the alarm office. Put in the last digit of the current calendar year and then put in the leading zeros. Examples: In 1988 the number 1 reads "8000001', while the number 10 would read "8000010".

**Date:**            Use two numbers for month, two numbers for day, and two numbers for year. For example: January 2, 1988 would be entered as [0][1][0][2][8][8].

**Incident Address:**            Record the address where the incident occurred, if different from the dispatch address, advise the Fire Alarm Division of the correct address.

           **Examples:**      (Two examples are given for each format.)

           **(1)**      Block Number
5680 SW 87 Avenue
1500 SW 104 Street

           (2)      Intersection
15 Street SW Redland Rd
9 Fiagier Street SW 105 Avenue

           Note: If there is not a number at the beginning of the address place a 9 in front of the street name.

           (3)      Exact Quadrant
6000 Galloway Road
1900 Ali Baba Avenue Room #5

           Note: Do not use periods after abbreviations.

           (4)      Landmark
9 MDCON Scott Hail
9 MIA LL ConE

           Note: If there is not a number at the beginning of the address place a 9 in front of the street name.

(5) Descriptive
9 West Arsenicker Key Shoals
5 Mi W Hmstd Gen Airport in field

Note:Use blanks between abbreviations, do not use slashes or periods. If a number is included in the beginning of the address do not enter a 9 (meaning no number) in front of the street address.

---

## PATIENT SECTION

**Patient's Address:** Put Last Name, (comma) First Name with no spaces before or after comma. If there is a title or a middle initial, leave a blank and add the title or initial after the first name.

**Patient #:** Each patient on every rescue incident should be assigned a patient number by the OIC. The patient number will follow the patient on every report. This is the patient number for the entire incident not just the patients involved in a water rescue.

**Patient Home Address:** Write in the patient's home address including the city state and zip aide.

**Age:** Patient's age to the nearest year. An infant less than one year would be entered as '01'.

**Sex:** Patient's sex. Enter M for male, F for female.

**Ethnic:** Enter the patient's ethnic background from Table 9..

**Dispatch:** Enter the time that the reporting unit was dispatched.

**Arrival:** Enter the time of your arrival.

**In Service:** Enter the time that the unit returned to service.

**Day:** Enter the code number for the day of the week from the listed table.

**Body Of Water:** Enter the code number from the report which identifies the type of body of water.

**Description On Incident:**
Enter the type of incident. if 'OTHER", write In what type.

**Water Conditions:** Write in feet of visibility and depth and speed of current in knots.

**Mode Of Operation:** Enter the method by which the operation was executed. For "OTHER', write a description.

**Consciousness:** Enter from the listed table, the level of consciousness on both ARRIVAL and RELEASE.

**Patient Time Underwater:**

Enter the estimated or reported time that the patient was underwater. Use two digits for hours and two digits for minutes. Example: One hour-twenty minutes would be entered as [0][l] [2][0].

**Air Pocket:**

indicate if there was an air pocket available to the patient.

**Area in Vehicle:**

Enter the location in the vehicle where the patient was found if applicable. For "OTHER', write a brief description.

**# Of Patients:**

Enter the actual number of patients involved in the water rescue portion of the incident. Enter "9" for 9 or more victims and indicate in the narrative the actual number if different.

Note: This is not to be confused with the TOTAL # (of patients) section on the Rescue Patient Record. These numbers may or may not be the same.

**Drowning By Catagory:**

Enter the category of drowning which applies. For "OTHER", write a brief description.

**Patient Position:**

Enter the code number which indicates the position in which the patient was found.

**Patient Dress:**

Enter, the code which indicates how the patient was dressed when recovered. For "OTHER', write a brief description.

**Injuries To Rescue Divers:**

Enter '1" for yes, "2" for no and describe in the narrative.

---

## SEARCH/NARRATIVE SECTION

**Search Pattern:**     Enter "1" for a linear search and "2" for a circular pattern.

**Datum Point Available:**  Enter "1" if a datum point was available and '2' if not.

**Victum Distance From Datum Point:**

Indicate in actual feet, how far from the datum point the victim was recovered. Use leading zeroes if needed. For example: If the victim was found 40' from the datum point enter- [0][4][0].

**Show Diagram:**

Draw a diagram of the incident showing all pertinent landmarks. (include datum point, location of victim(s), roadways, etc.)

**( ) Additional Narrative:**

Check the box to indicate an additional narrative record has been completed. A SUPPLEMENTAL NARRATIVE/UNIT RECORD will be used when the narrative space provided is inadequate to record pertinent information.

**Print:**

List the rescue divers and line tenders. The person reporting signs next to their name.

3:44

RECORD

**0 7**

1  2  10/91

# WATER RESCUE RECORD
### FIRE RESCUE INCIDENT REPORT DADE COUNTY, FLORIDA

UNIT #   SHIFT

A
B
C

3   5   6

DEPT   INCIDENT NUMBER   M M D D Y Y   PATIENT HOME ADDRESS

7  8   9   15   16   21

PATIENT'S LAST NAME (comma)   FIRST NAME (space)   MIDDLE INITIAL   PATIENT NUMBER

22   47   48

DATE OF BIRTH   INCIDENT ADDRESS
M M D D Y Y

49   54   55   77

**SEX**
M
F
78

**ETHNIC**
1 WHITE
2 BLACK
3 SPANISH
**SEE TABLE ( )
79

DISPATCH   ARRIVAL   IN SERVICE   DAY
1 2 3 4 5 6 7
S M T W R F S

80   83   84   87   85   91   92

**BODY OF WATER**
1 ROCK PIT  4 OCEAN
2 CANAL     5 RIVER
3 BAY       6 POOL
93   7 OTHER _____

**DESCRIPTION OF INCIDENT**
1 SURFACE RESCUE OF DROWNING VICTIM
2 SEARCH & RECOVERY OF DROWNING VICTIM
3 EXTRICATION FROM CAR
94   4 OTHER _____

**WATER CONDITIONS**
VISIBILITY _____ FEET
DEPTH _____ FEET
CURRENT _____ KNOTS

**MODE OF OPERATION**
1 BREATH HOLD
2 SCUBA
3 OTHER _____
95

**CONSCIOUSNESS**
ARRIVAL   RELEASE
1 CON
2 DRO
3 UNC
96   4 NVS   97

**PATIENT TIME UNDERWATER**
HOURS   MINUTES
98   99   100   101

**AIR POCKET**
1 YES
2 NO
102

**AREA IN VEHICLE**
1 LEFT FRONT    4 LEFT REAR
2 RIGHT FRONT   5 RIGHT REAR
3 CENTER FRONT  6 CENTER REAR
103   7 OTHER _____

**# OF PATIENTS**
**DESCRIBE IN NARRATIVE
194

**DROWNING BY CATEGORY**
1 ACCIDENT  5 BOAT RELATED
2 SUICIDE   6 AIRCRAFT
3 HOMICIDE  7 OTHER
105   4 AUTO RELATED _____

**PATIENT POSITION**
1 FLOATING
2 ON BOTTOM FACE UP
3 ON BOTTOM FACE DOWN
106   4 IN VEHICLE

**PATIENT DRESS**
1 STREET CLOTHES 4 SCUBA DIVING GEAR
2 SWIM SUIT      5 OTHER
3 SKIN DIVING GEAR _____
107

## SEARCH INFORMATION

**INJURIES TO RESCUE DIVERS**
1 YES (IF YES, DESCRIBE ON SUPPLEMENTAL
2 NO           NARRATIVE)
108

**SEARCH PATTERN**
1 LINEAR
2 CIRCULAR
109

**DATUM POINT AVAILABLE**
1 YES
2 NO
110

**VICTIM DISTANCE FROM DATUM POINT**
FEET
111   113

SHOW DIAGRAM & INCLUDE THE FOLLOWING: _____ (1) SHORELINE  (2) ACCESS ROAD  (3) DIRECTION OF NORTH  (4) AREA OF PROBABILITY
(5) DATUM POINT, IF APPLICABLE  (6) ACTUAL SEARCH AREA  (7) LOCATION OF PATIENT  (8) PERTINENT TERRAIN FEATURES & ENVIRONMENTAL FACTORS
(9) LOCATE PATIENT AND OTHER OBJEFTS BY DIMENSIONS IN FEET

VARIATIONS OF SEARCH PATTERN OR SOLUTIONS TO PROBLEMS ENCOUNTERED:

_____

_____

_____

( ) NARRATIVE ATTACHMENT

| LIMITATION & PROBLEMS OF SEARCH: | PRINT (LAST NAME, FIRST NAME) |
|---|---|
|  | RESCUE DIVER |
|  | RESCUE DIVER |
|  | LINE TENDER |
|  | LINE TENDER                    DATE |

**DISTRIBUTION:**   WHITE — ORIGINAL   YELLOW — WATER RESCUE   PINK — MIS   GOLD — STATION

125.01-132 Rev. 10/91

# APPENDIX C: TECHNICAL RESCUE RESOURCE LIST

## ORGANIZATIONS, AGENCIES AND ASSOCIATIONS THAT CAN PROVIDE TECHNICAL RESCUE ASSISTANCE

**California State Office of Emergency Services**
Urban Search and Rescue Program
2151 East D Street, Suite 203A
Ontario, CA 91764

**Federal Emergency Management Agency**
Public Assistance Division
Office of Disaster Assistance Programs
SL-OE-FR-OP
500 C Street, SW.
Washington, D.C. 20472
(202) 646-2442

**International Association of Dive Rescue Specialists, Inc.**
201 North Link Lane
P.O. Box 5259
San Clemente, CA 92674-5259
(714) 489-2004

**International Association of Fire Chiefs**
Urban Rescue and Structural Collapse Committee
4025 Fair Ridge Drive, Suite 300
Fairfax, VA 22033-2868
**(703) 273-0911**

**Mountain Rescue Association**
2144 South 1100, Suite 150-375
Salt Lake City, UT 84106
(801) 328-0523

**National Association for Search and Rescue**
4500 Southgate Place, Suite 100
Chantilly, VA 22021
(703) 222-6277

**National Fire Protection Association**
Standards Branch
P.O. Box 9101
1 Batterymarch Park
Quincy, MA 02269-9101
(617) 770-3000

**National Institute for Urban Search and Rescue**
P.O. Box 91648
Santa Barbara, CA 93190-1648

**Plan Bulldozer Committee**
Greater Kansas City Metropolitan Region
Ed DeSoignie
Heavy Constructors Association
3101 Broadway, Suite 780
Kansas City, MO 6411
(816) 753-6443

**Urban Search and Rescue Inc.**
P.O. Box 2570
Camarillo, CA 93011-2570

**U.S. Fire Administration**
Fire Technical Programs
16825 South Seton Avenue
Emmitsburg, MD 21727

## PUBLICATIONS

The U.S. Fire Administration has a host of publications that are of great assistance when starting a technical rescue team. Most publications are free and can be ordered by writing the U.S. Fire Administration Publications Center, 16825 South Seton Avenue, Emmitsburg, MD 21727. When ordering, be sure to note the document number which appears in parentheses below.

- *Technical Rescue (US&R) Incident Investigations* (different technical rescue incidents and lessons learned are discussed) (FA120-FA125)

- *Technical Rescue Technology Assessment* (FA-135)

- *New Technologies in Vehicle Extrication* (FA-152)

- *Guide to Funding Alternatives* for *Fire and EMS Departments* (FA-141)

- *Guide to Developing and Managing an Emergency Service Infection Control Program* (FA-112)

- *Protective Clothing and Equipment Needs* of *Emergency Responders for Urban Search and Rescue Missions* (FA-136)

- *Emergency Medical Services Management Resources Directory* (FA-119)

- *Minimum Standards* of *Structural Firefighting Protective Clothing and Equipment* (FA-137)

- *Major Fires Report Series* (dozens of major fire incidents and lessons learned from across the country are discussed) (ask for order form)

### Other Documents

*Personal Property Utilization and Disposal Guide*
**General Services Administration**
Centralized Mailing List Service
P.O. Box 6477
Fort Worth, TX 76115

*How to Buy Surplus Personal Property (from the United States Department of Defense)*
**Defense Reutilization  and Marketing Service**
National Sales Office
P.O. Box 5275 DDRC
*2163* Airways Boulevard
Memphis, TN 38114

*Water Rescue Program Manual*
**Metro-Dade County Fire Department**
Special Operations Division
6000 S. W. 87th Avenue
Miami, FL 33173-1698
*(305) 596-8538*

*Fire Command*
**Alan Brunacini, Author**
National Fire Protection Association
1 Battermarch Park
Quincy, MA 02269-9101
(617) 770-3000

## REGULATIONS AND STANDARDS

Some of the organizations which have developed regulations or standards guiding technical rescue operations and/or training are listed below.

**ASTM**
1916 Race Street
Philadelphia, PA

**National Fire Protection Association**
Standards Branch
P.O. Box 9101
1 Batterymarch Park
Quincy, MA 02269-9101
(617) 770-3000
*(The NFPA Liaison to the Committee on Technical Rescue is Charles Smeby.)*

**Occupational Safety and Health Administration (OSHA)**
Attention Mr. Thomas Seymour
U.S. Department of Labor
Safety Standards Program
200 Constitution Avenue, NW
Washington, D.C. 20210
*Copies of OSHA regulations are usually available from a local library)*

**National Association for Search and Rescue**
*Guidelines for Public Safety Diving*
4500 Southgate Place, Suite 100
Chantilly & VA 22021
(703) 222-6277

## Other U.S. Government Publications

Superintendent of Documents
U.S. Government Printing Office
Washington, D.C.   20402

## TRAINING RESOURCES

### Public Training Resources

Several states and localities have begun to develop rescue training curriculum. Below is a list of some of these:

*Rescue Systems I and II*
California State Fire Marshal
California Fire Service Training and Education System
7171 Bowling Drive, Suite 600
Sacramento, CA 95823

Indianapolis Fire Department
*Basic Emergency Rescue Technician (BERT) Training Program*
555 North New Jersey Street
Indianapolis, IN 46204

Montgomery County (MD) Department of Fire and Rescue
*Practical Rescue Course*
Fire and Rescue Training Academy
10025   Darnestown Road
Rockville, MD 20850

Virginia Office of Fire Programs
P.O. Box 47
Orange, VA   22960

### Private Training Resources

There are many private companies which have developed curriculum and delivered training for technical rescue. Some of these are listed below. This list is only a partial list of training firms. Other training resources can be found in fire and rescue trade publications.

*Rescue 3 International*
(800) 457-3728

*ROCO Rescue Corporation*
(904) 265-0525

*L.A. Emergency Management and Training*
(315) 453-3630

*Dive Rescue International*
(303) 482-0887

# APPENDIX D: TECHNICAL RESCUE EQUIPMENT LIST

# Technical Rescue Equipment list

This section contains a listing of equipment you may consider acquiring for a technical rescue team. The list is designed to assist any agency or organization in identifying general tools, supplies, and equipment necessary to safely and effectively undertake technical rescue operations. The equipment listing is designed to address the following tactical capabilities:

- Rope/High Angle Rescue Operations
- Trench Rescue Operations
- Structural Collapse Rescue Operations
- Confined Space Rescue Operations
- Industrial/Agricultural Rescue Operations
- Water/Ice Rescue Operations

The equipment listing is divided into subcategories to address different rescue operational and training levels discussed in this manual. The subcategories include:

- First Responder (Awareness Level) Rescue Team
- Operations Level Rescue Team
- Advanced Heavy Rescue Team
- Advanced Heavy Rescue Team - Tool Box Inventory
- Advanced Heavy Rescue Team - Trench/Structural Shoring Unit Inventory
- First Responder/Operations Level Water Rescue Team
- Advanced Level Water Rescue Team

The First Responder Rescue Team section identifies basic tools and equipment (mostly hand tools) that personnel trained to the awareness level can use to undertake basic technical rescue operations. This is equipment that would typically be carried on any first line fire apparatus such as an engine company.

The Operations Level Rescue Team section lists basic tools and equipment (hand and basic power tools) that personnel trained to the operations level can use to undertake basic to moderate rescue operations. This is equipment that would typically be carried on any first line fire apparatus or light rescue squad.

The Advanced Heavy Rescue Team section identifies tools and equipment (hand tools, basic and specialized power tools) that personnel trained to advanced levels (operations and technician level) can use to undertake basic through complex rescue operations. This is equipment that may already be present on a heavy rescue squad or specialized rescue unit. Other items listed are specialized pieces of equipment that require moderate to significant funding to procure. A short list is included at the end of quite specialized equipment that some advanced teams are currently using. The purchase of these items may be deferred or delayed until funding and associated training are available.

In addition, a separate Tool Box Inventory section which identifies the numerous hand tools that may be needed for Heavy Rescue Teams is provided.

The Trench/Structural Shoring Unit inventory section lists equipment that is required predominantly for trench and structural collapse operations. Due to the size, weight, and volume required to transport this type of equipment, most technical rescue teams relegate this equipment to a separate vehicle or unit that may not respond on every technical rescue dispatch.

The First Responder/Operations Level Water Rescue Team section identifies basic tools and equipment for an awareness to operations level water rescue team. These tools will allow basic rescues to be performed from shore without entry by personnel. This section is divided into equipment for swiftwater, flatwater, or ice rescue operations.

The Advanced Level Water Rescue Team section lists basic and advanced tools that an advanced team may need for shore-based or entry rescues. This section is also divided into equipment for swiftwater, flatwater, or ice rescue operations.

**Technical Rescue Equipment**

Discipline:
1 - Rope/High Angle Operations
2 - Trench Operations
3 - Structural Collapse Operations
4 - Confined Space Operations
5 - Industrial/Agricultural Ops

## FIRST RESPONDER (AWARENESS LEVEL) RESCUE TEAM

| Quantity | Item | Discipline | Unit Price ($) | cost ($) |
|---|---|---|---|---|
| 2 | Axe, pick head | All | 32 | 64 |
| 2 | Axe, flat head | All | 26 | 52 |
| 2 | Banner guard tape, "Fireline," roll | All | 17 | 34 |
| 2 | Body bags | All | 34 | 68 |
| 4 | Carabiners, large, steel, locking "D" | All | 18 | 72 |
| 1 | Cord, electric, 200 ft: w/ plugs/adapters | All | 130 | 130 |
| 4 | Descenders with ears, large, steel | All | 28 | 112 |
| 1 | Electrical testing device (amprobe, volt/ohm meter) | All | 75 | 75 |
| 1 | Extinguisher, dry chemical | All | 87 | 87 |
| 1 | Extinguisher, CO | All | 246 | 246 |
| 1 | Extinguisher, water, 2 1/2 gal. | All | 69 | 69 |
| 1 | First aid kit | All | 165 | 165 |
| 1 | Generator, gas-powered, 5000W | All | 1,370 | 1,370 |
| 4 | Goggles, safety | All | 6 | 24 |
| 1 | Hacksaw, carbide blade | All | 26 | 26 |
| 1 | Halligan tool | All | 127 | 127 |
| 2 | Hammer, carpenters, 22 oz. | All | 25 | 50 |
| 4 | Handlight, battery-operated | All | 75 | 300 |
| 4 | Hearing protection headsets | All | 18 | 72 |
| 1 | Ladder, roof, 14 foot | All | 240 | 240 |
| 1 | Ladder, attic, 10 foot | All | 200 | 200 |

**Technical Rescue Equipment**

Discipline:
1 - Rope/High Angle Operations
2 - Trench Operations
3 - Structural Collapse Operations
4 - Confined Space Operations
5 - Industrial/Agricultural Ops

## FIRST RESPONDER (AWARENESS LEVEL) RESCUE TEAM

| Quantity | Item | Discipline | Unit Price ($) | cost ($) |
|---|---|---|---|---|
| 1 | Ladder, extension, 24 foot | All | 450 | 450 |
| 1 | Latex gloves, box | All | 8 | 8 |
| 2 | Light, Circle D 500W | All | 116 | 232 |
| 1 | Mallet, rubber | All | 32 | 32 |
| 1 | Masks, dust, box | All | 9 | 9 |
| 1 | Oxygen/suction unit, portable | All | 510 | 510 |
| 1 | Pike pole, 6 foot | All | 43 | 43 |
| 1 | Rope, utility, 100 foot, ½" | All | 35 | 35 |
| 1 | Rope, lifeline, 150 foot, 12.7MM static Kemmantle ( with rope bag) | ALL | 110 | 220 |
| 1 | salvage cover, 12 X 18 | ALL | 189 | 189 |
| 1 | Smoke ejector, electric, 16" | ALL | 429 | 429 |
| 1 | Tool box and small mechanics tools set | ALL | 150 | 150 |
| 4 | Webbing, tublar, 2" wide, 20 foot length | ALL | 14 | 56 |
| 1 | Bolt cutter, 18" | 3,4,5 | 38 | 38 |
| 1 | Crow bar, 24" | 3,4,5 | 12 | 12 |
| 1 | K-Tool, lock removal device | 3,4,5 | 81 | 81 |
| 4 | Personal alert device (PASS) | 3,4,5 | 498 | 1,992 |
| 1 | Pry bar | 3,4,5 | 34 | 34 |
| 4 | Self-contained breathing apparatus (SCBA) | 3,4,5 | 1,845 Text | 7,380 |
| 2 | Wheel chock, metal (fulcrum/crib) | 3,4,5 | 21 | 42 |
| 1 | Key tool, water utility | 2,3,4,5 | 12 | 12 |

**Technical Rescue Equipment**

Discipline:
1 - Rope/High Angle Operations
2 - Trench Operations
3 - Structural Collapse Operations
4 - Confined Space Operations
5 - Industrial/Agricultural Ops

## FIRST RESPONDER (AWARENESS LEVEL) RESCUE TEAM

| Quantity | Item | Discipline | Unit Price ($) | cost ($) |
|----------|------|------------|----------------|----------|
| 1 | Key tool, gas utility | 2, 3, 4, 5 | 12 | 12 |
| 2 | Shovel, dirt, round point | 2, 3, 4, 5 | 37 | 74 |

# Technical Rescue Equipment

Discipline:
1 - Rope/High Angle Operations
2 - Trench Operations

3 - Structural Collapse Operations
4 - Confined Space Operations
5 - Industrial/Agricultural Ops

## OPERATIONS LEVEL RESCUE TEAM

| Quantity | Item | Discipline | Unit Price ($) | cost ($) |
|---|---|---|---|---|
| 2 | Axe, pick head | All | 32 | 64 |
| 2 | Axe, flat head | All | 26 | 52 |
| 1 | Backboard | All | 32 | 32 |
| 2 | Body bags | All | 34 | 68 |
| 2 | Crow bar, 36" carpenter | All | 16 | 32 |
| 2 | Debris bag | All | 35 | 70 |
| 1 | Electrical tesing device (Amprobe, volt/ohm meter) | All | 75 | 75 |
| 2 | Extinguisher, dry chemical | All | 87 | 174 |
| 2 | Extinguisher, pressurized water, 2½ gal. | All | 42 | 84 |
| 2 | Extinguisher, CO | All | 246 | 492 |
| 2 | "Fireline" banner guard, roll | All | 17 | 34 |
| 1 | First aid kit | All | 165 | 165 |
| 1 | Generator, gas-powered, 7500W or greater | All | 1,840 | 1,840 |
| 4 | Goggles, safety | All | 6 | 24 |
| 4 | Hammer, carpenter, 22 oz. | All | 25 | 100 |
| 4 | Handlight, battery-operated | All | 75 | 300 |
| 1 | Harness cable set for Stokes | All | 80 | 80 |
| 4 | Hearing protection headsets | All | 18 | 72 |
| 1 | Ladder, attic, 10 foot | All | 200 | 200 |
| 2 | Ladder, extension, 35 foot | All | 1,340 | 2,680 |
| 1 | Ladder, folding, Little Giant | All | 480 | 480 |

**Technical Rescue Equipment**

Discipline:
1 - Rope/High Angle Operations
2 - Trench Operations
3 - Structural Collapse Operations
4- Confined Space Operations
5 - Industrial/Agricultural Ops

## OPERATIONS LEVEL RESCUE TEAM

| Quantity | Item | Discipline | Unit Price ($) | cost ($) |
|---|---|---|---|---|
| 1 | Ladder, roof, 14 foot | All | 285 | 285 |
| 2 | Ladder, extension, 28 foot | All | 620 | 1,240 |
| 1 | Latex gloves, box | All | 8 | 8 |
| 2 | Light, Circle D 500W | All | 116 | 232 |
| 1 | Mallet, rubber | All | 28 | 28 |
| 1 | Masks, dust, box | All | 9 | 9 |
| 1 | Oxygen/suction unit, portable | All | 510 | 510 |
| 2 | Pry bar | All | 34 | 68 |
| 1 | Reeves stretcher | All | 197 | 197 |
| 3 | Rope, lifeline, 150 foot, 12.7MM static Kemmantle with rope bag) | All | 110 | 330 |
| 2 | Rope, 20 ft. section of 12.7MM Kemmantle | All | 19 | 38 |
| 1 | Rope, utility, 100 foot, %" | All | 35 | 35 |
| 3 | Salvage cover, 12 X 18 | All | 189 | 567 |
| 2 | Salvage cover, 14 X 18 | All | 211 | 422 |
| 1 | SKED stretcher | All | 300 | 300 |
| 2 | Sledge hammer, 12 lb., long handle | All | 32 | 64 |
| 2 | Sledge hammer, 8 lb., short handle | All | 28 | 56 |
| 1 | Stokes basket | All | 290 | 290 |
| 2 | Tag lines, 100 ft. | All | 14 | 28 |
| 4 | Tape, duct, roll | All | 3 | 12 |
| 1 | Tool box and mechanics tools set | All | 400 | 400 |

**Technical Rescue Equipment**

Discipline:
1 - Rope/High Angle Operations
2 - Trench Operations
3 - Structural Collapse Operations
4 - Confined Space Operations
5 - Industrial/Agricultural Ops

## OPERATIONS LEVEL RESCUE TEAM

| Quantity | Item | Discipline | Unit Price ($) | Cost ($) |
|---|---|---|---|---|
| 4 | Personal alert device (PASS) | 4 | 498 | 1,992 |
| 4 | SCBA bottles, extra | 4 | 365 | 1,460 |
| 4 | Self-contained breathing apparatus (SCBA) | 4 | 1,845 | 7,380 |
| 1 | Line throwing gun, w/ accessories | 1, 2 | 365 | 365 |
| 8 | Carabiner, large, steel, locking "D" | 1, 4 | 18 | 144 |
| 4 | Descender with ears, large, steel | 1, 4 | 28 | 112 |
| 4 | Life belts | 1, 4 | 127 | 508 |
| 4 | Webbing, tubular, 2" tide, 20 foot length | 1, 4 | 14 | 56 |
| 2 | Gas cans ( gasoline and gas/oil mix) | 2, 3 | 31 | 62 |
| 1 | Handsaw, cross cut | 2, 3 | 31 | 31 |
| 1 | Nails, 16P, double-head, box | 2, 3 | 50 | 50 |
| 1 | Saw, chain, gasoline-powered, 16" | 2, 3 | 360 | 360 |
| 2 | Saw, circular, gasoline-powered, 16" w/ carbide, metal and masonry blades | 2, 3 | 870 | 1,740 |
| 1 | Saw, circular skilsaw, electric | 2, 3 | 180 | 180 |
| 1 | Saw, chain, gasoline-powered, 24" | 2, 3 | 490 | 490 |
| 1 | Saw, reciprocating, electric | 2, 3 | 195 | 195 |
| 1 | Axe, crash | 3, 4, 5 | 44 | 44 |
| 1 | Axe, pry | 3, 4, 5 | 35 | 35 |
| 1 | Bolt cutter, 18" | 3, 4, 5 | 32 | 32 |
| 1 | Bolt cutter, 36" | 3, 4, 5 | 81 | 81 |
| 1 | Closet hook | 3, 4, 5 | 69 | 69 |

**Technical Rescue Equipment**

Discipline:
1 - Rope/High Angle Operations
2 - Trench Operations

3 - Structural Collapse Operations
4 - Confined Space Operations
5 - Industrial/Agricultural Ops

## OPERATIONS LEVEL RESCUE TEAM

| Quantity | Item | Discipline | Unit Price ($) | cost ($) |
|---|---|---|---|---|
| 2 | Halligan tool | 3, 4, 5 | 127 | 254 |
| 1 | K-Tool, lock removal device | 3, 4, 5 | 81 | 81 |
| 1 | Poling tool, elevator | 3, 4, 5 | 65 | 65 |
| 2 | Wheel chock, metal (fulcrum/crib) | 3, 4, 5 | 21 | 42 |
| 1 | Atmospheric monitor (GasTrac, MSA Explosion Meter, etc.) | 2, 3, 4, 5 | 440 | 440 |
| 2 | Bar, chin-up (for smoke ejectors) | 2, 3, 4, 5 | 13 | 26 |
| 1 | Blower, gasoline-powered, 20" | 2, 3, 4, 5 | 2,500 | 2,500 |
| 1 | Broom, street | 2, 3, 4, 5 | 13 | 13 |
| 1 | Key tool, gas utility | 2, 3, 4, 5 | 12 | 12 |
| 1 | Key tool, water utility | 2, 3, 4, 5 | 12 | 12 |
| 1 | Pike pole, 12 foot | 2, 3, 4, 5 | 59 | 59 |
| 1 | Pike pole, 10 foot | 2, 3, 4, 5 | 41 | 41 |
| 1 | Pike pole, 14 foot | 2, 3, 4, 5 | 66 | 66 |
| 1 | Pike pole, 16 foot | 2, 3, 4, 5 | 70 | 70 |
| 1 | Pipe wrench, 18" | 2, 3, 4, 5 | 45 | 45 |
| 2 | Pitch forks | 2, 3, 4, 5 | 47 | 94 |
| 1 | Porto-power hydraulic tool, 4-ton | 2, 3, 4, 5 | 249 | 249 |
| 1 | Rabbit hydraulic forcible entry tool | 2, 3, 4, 5 | 2,040 | 2,040 |
| 4 | Respirator, full face w/ HEPA filters | 2, 3, 4, 5 | 197 | 788 |
| 2 | Shovel, dirt, round point | 2, 3, 4, 5 | 37 | 74 |
| 2 | Shovel, scoop | 2, 3, 4, 5 | 32 | 64 |

**Technical Rescue Equipment**

Discipline:
1 - Rope/High Angle Operations
2 - Trench Operations

3 - Structural Collapse Operations
4 - Confined Space Operations
5 - Industrial/Agricultural Ops

## ADVANCED HEAVY RESCUE TEAM

| Quantity | Item | Discipline | Unit Price ($) | Cost ($) |
|---|---|---|---|---|
| 10 | Carpet squares, 2X2 | All | 3 | 30 |
| 2 | Chain, Grade 7, $^3/_8$", 20 ft. w/ hooks | All | 40 | 80 |
| 1 | Come-along, 4-ton | All | 126 | 126 |
| 1 | Come-along, 1 -ton | All | 94 | 94 |
| 600 | Cord, electric, 12G., three-wire grounded (various lengths), w/ adapters | All | 420 | 420 |
| 1 | Edge frame (portable frame for rope raising operations) | All | 1,690 | 1,690 |
| 1 | Forcible entry ram, sliding, kit | All | 510 | 510 |
| 2 | Gas can (gasoline and gas/oil mix) | All | 14 | 28 |
| 1 | Generator, 20KW or greater, vehicle-mounted | All | 7,180 | 7,180 |
| 1 | Hailing device, battery-powered (bullhorn) | All | 165 | 165 |
| 2 | Jack-o-lanterns (cord reel/light set) | All | 209 | 418 |
| 4 | Junction boxes, electric | All | 85 | 340 |
| 4 | Lights, Circle D or 500W halogens | All | 116 | 464 |
| 1 | Line throwing gun w/ accessories | All | 390 | 390 |
| 1 | Oxygen manifold "Multilator" w/ $O_2$ bottles | All | 790 | 790 |
| 4 | Rope, lifeline, 150 ft, 12.5MM kemmantle, static, w/ bag | All | 110 | 440 |
| 4 | Sheave, cable, 4" | All | 65 | 260 |
| 6 | Stakes, steel | All | 4 | 24 |
| 1 | Stretcher, SKED-type | All | 447 | 447 |
| 1 | Tripod, 8 ft. | All | 690 | 690 |

# Technical Rescue Equipment

Discipline:
1 - Rope/High Angle Operations
2- Trench Operations
3 - Structural Collapse Operations
4 - Confined Space Operations
5 - Industrial/Agricultural Ops

## ADVANCED HEAVY RESCUE TEAM

| Quantity | Item | Discipline | Unit Price ($) | cost ($) |
|---|---|---|---|---|
| 4 | Tripod/light stands, telescoping | All | 460 | 1,840 |
| 1 | Winch, electric, 4-ton, vehicle mounted | All | 992 | 992 |
| 1 | Hand winch for tripod, w/ 100 ft. cable | 4 | 1,207 | 1,207 |
| 4 | Head lamps, low voltage, intrensically safe | 4 | 728 | 2,912 |
| 6 | SCBA bottle, extra | 4 | 365 | 2,190 |
| 6 | Self-contained breathing apparatus (SCBA) | 4 | 1,845 | 11,070 |
| 2 | Supplied air breathing system w/ manifold, regulator, 600' air line, 2 face pieces, escape bottles, and accessories (one system supplies two entry personnel) | 4 | 2,600 | 5,200 |
| 4 | Ascender, aluminum, 1/2" | 1,4 | 39 | 156 |
| 24 | Carabiner, large, steel, locking "D" | 1,4 | 18 | 432 |
| 1 | Cord, prusik, 8MM, 200 ft. | 1,4 | 90 | 90 |
| 2 | Daisy chain, 5 ft. | 1,4 | 14 | 28 |
| 12 | Descender, w/ ears, large, steel | 1,4 | 28 | 336 |
| 2 | Edge roller | 1,4 | 118 | 236 |
| 2 | Etrier (3 and 4 step) | 1,4 | 19 | 38 |
| 4 | Harness, personal, Class II | 1,4 | 89 | 356 |
| 2 | Harness, Class III | 1,4 | 150 | 300 |
| 6 | Prusi k, long | 1,4 | 3 | 18 |
| 6 | Prusi k, short | 1,4 | 2 | 12 |
| 10 | Pulley, single | 1,4 | 25 | 250 |
| 4 | Pulley, single, 2" X 1/2" | 1,4 | 41 | 164 |

**Technical Rescue Equipment**

Discipline:
1 - Rope/High Angle Operations
2 - Trench Operations
3 - Structural Collapse Operations
4 - Confined Space Operations
5 - Industrial/Agricultural Ops

## ADVANCED HEAVY RESCUE TEAM

| Quantity | Item | Discipline | Unit Price ($) | Cost ($) |
|---|---|---|---|---|
| 2 | Pulley, double, 2" X 1/2" | 1, 4 | 65 | 130 |
| 1 | Rope, 12.7MM kemmantle, 300 ft., static | 1, 4 | 240 | 240 |
| 2 | Webbing, 2", 30" | 1, 4 | 1 | 2 |
| 3 | Webbing, 2", 20 ft. | 1, 4 | 3 | 9 |
| 4 | Whistles | 1, 4 | 3 | 12 |
| 1 | Air bag, high pressure, 136-ton set, includes: l-ton bag, 3-ton bag, 5-ton bag, 12-ton bag, 17-ton bag, 22-ton bag, 32-ton bag, 44-ton bag | 2, 3 | 4,970 | 4,970 |
| 1 | Air bag high pressure regulator/control valve kit | 2, 3 | 576 | 576 |
| 1 set | Air bag, low pressure, small; medium, large | 2, 3 | 1,800 | 1,800 |
| 30 | Cribbing, 2"X4"X2 ft. | 2, 3 | 1 | 30 |
| 20 | Cribbing, 4"X4"X3 ft. | 2, 3 | 2 | 40 |
| 2 | Jack, Hi-Lift, 48" | 2, 3 | 85 | 170 |
| 1 | Nail/fastener gun w/ accessories, charges, fasteners, nails (Hilti, Paslode, etc.) | 2, 3 | 740 | 740 |
| 1 | Pneumatic airgun breaker tool | 2, 3 | 2,245 | 2,245 |
| 2 | Saw, chain, 16", gasoline, w/ carbide chain | 2, 3 | 893 | 1,786 |
| 2 | Saw, chain, 14" electric | 2, 3 | 280 | 560 |
| 1 | Saw, worm gear | 2, 3 | 374 | 374 |
| 20 | Wedges, 4"X4"X2 ft. | 2, 3 | 1 | 20 |
| 1 | Post hole digger | 1, 2, 3 | 40 | 40 |

# Technical Rescue Equipment

## ADVANCED HEAVY RESCUE TEAM

| Quantity | Item | Discipline | Unit Price ($) | Cost ($) |
|---|---|---|---|---|
| 1 | Atmospheric monitor, oxygen analyzer-type, personal | 2, 3, 4 | 2,000 | 2,000 |
| 1 | Atmospheric monitor, gas detector | 2, 3, 4 | 560 | 560 |
| 1 | Atmospheric monitor, Explosion meter | 2, 3, 4 | 440 | 440 |
| 2 | Personal atmospheric monitors (Industrial Scientific HMX-271 type) | 2, 3, 4 | 1,700 | 3,400 |
| 24 | Respirator cartridges (asbestos rated) | 2, 3, 4 | 110 | 2,640 |
| 6 | Respirators, full face w/ cartridge attachments | 2, 3, 4 | 197 | 1,182 |
| 1 | Ventilation blower, electric, utility-type, w/ 15' tubing | 2, 3, 4 | 547 | 547 |
| 1 | Hydraulic spreader/jaws/cutter, multi-purpose (Hurst-type) | 2, 3, 5 | 3,500 | 3,500 |
| 1 | Hydraulic tool power unit (Hurst-type) w/ hoses and accessories | 2, 3, 5 | 2,800 | 2,800 |
| 2 | Jack, hydraulic, 20-ton, bottle-type | 2, 3, 5 | 150 | 300 |
| 2 | Jack, hydraulic, 10-ton, bottle-type | 2, 3, 5 | 85 | 170 |
| set | Jimmi Jack stabilization system w/2 short and 2 long stabilizers and accessories<br>accessories   2 - point tips<br>2 - wedges   2 - chains<br>2 - V-shape tips<br>2 - J-hooks   4 - swivels<br>1 - regulator | 2, 3, 5 | 7,800 | 7,800 |
| 2 | Rams (short, long) | 2, 3, 5 | 1,700 | 3,400 |
| 1 | Bolt cutter, 36" | 3, 4, 5 | 62 | 62 |
| 1 | Bolt cutter, 24" | 3, 4, 5 | 49 | 49 |
| 1 | Cut-off tool, pneumatic (Wizzer saw) | 3, 4, 5 | 87 | 87 |
| 1 | Cutting torch set, oxy-acetylene | 3, 4, 5 | 430 | 430 |

**Technical Rescue Equipment**

Discipline:
1 - Rope/High Angle Operations
2 - Trench Operations
3 - Structural Collapse Operations
4 - Confined Space Operations
5 - Industrial/Agricultural Ops

## ADVANCED HEAVY RESCUE TEAM

| Quantity | Item | Discipline | Unit Price ($) | cost ($) |
|---|---|---|---|---|
| 1 | Exothermic cutting torch (Arcair, etc.) w/ accessories | 3, 4, 5 | 1,450 | 1,45C |
| 1 | Porto-power tool kit , lO-ton | 3, 4, 5 | 363 | 362 |
| 1 | Air hose/reel, 100 ft. | 2, 3, 4, 5 | 290 | 29c |
| 1 | Air compressor, vehicle-mounted | 2, 3, 4, 5 | 43 8 | 438 |
| 1 | Air chisel w/ blades | 2, 3, 4, 5 | 65 | 65 |
| 4 | Chocks, step-type | 2, 3, 4, 5 | 45 | 180 |
| 1 | Drill/driver," cordless | 2, 3, 4, 5 | 280 | 280 |
| 1 | Drill, 1/2 electric | 2, 3,4, 5 | 226 | 226 |
| 1 | Impact wrench, pneumatic | 2, 3, 4, 5 | 105 | 105 |
| 1 | Index set, drill bits | 2, 3, 4, 5 | 65 | 66 |
| 2 | **Saw,** reciprocating, electric | 2,3, 4, 5 | 192 | 384 |
| 1 | **Saw, worm** drive, 7%" | 2, 3, 4, 5 | 372 | 372 |
| 2 | Saw, rotary disk. 16" | 2, 3, 4, 5 | 870 | 1,740 |
| | | | | |
| | The following sophisticated equipment, while expensive, might be considered as latter purchases to augment and improve capabilities | | | |
| | | | | |
| 1 | Winch, portable, chain sawengine tpe | All | 2,800 | 2,800 |
| 1 | "US&R" concrete breaking kit (Stanley-type), includes:<br>  1 - power unit  1 - regular chain saw<br>  1 - diamond segmented concrete cutting chain saw<br>  3 - impact breakers (small, medium, large) | 2, 3 | 17,084 | 17,084 |

# Technical Rescue Equipment

Discipline:
1 - Rope/High Angle Operations
2 - Trench Operations
3 - Structural Collapse Operations
4 - Confined Space Operations
5 - Industrial/Agricultural Ops

## ADVANCED HEAVY RESCUE TEAM

| Quantity | Item | Discipline | Unit Price ($) | cost ($) |
|---|---|---|---|---|
| 1 | Drill/breaker tool, gasoline driven w/ accessories (Cobra Berema-type) | 2, 3 | 4,950 | 4,950 |
| 1 | Demolition hammer (Bosch-type) | 2, 3, 5 | 1,600 | 1,600 |
| 1 | Rebar cutter, electric, 3/4" capacity | 2, 3, 5 | 1,800 | 1,800 |
| 1 | Acoustic/seismic listening search device | 2, 3, 4, 5 | 8,000 | 8,000 |
| 1 | Hammer drill, 1/2" capacity (Hilti-type) | 2, 3, 4, 5 | 1,870 | 1,870 |
| 1 | Optical search device (fibre optic or Searchcam-type) | 2, 3, 4, 5 | 11,000 | 11,000 |

**Technical Rescue Equipment**

Discipline:
1 - Rope/High Angle Operations
2 - Trench Operations

3 - Structural Collapse Operations
4 - Confined Space Operations
5 - Industrial/Agricultural Ops

## ADVANCED HEAVY RESCUE TEAM - TOOL BOX INVENTORY

| Quantity | Item | Discipline | Unit Price ($) | cost ($) |
|---|---|---|---|---|
| 2 | Allen key set (US standard and metic) | All | | |
| 1 | Allen key set, folding, $^1/16$ " – $^3/16$" | All | | |
| 1 | Chain saw adjustment tool | All | | |
| 1 | Chisel, cold, $^3/4$" | All | | |
| 1 | Drive extension, 1/2" X 3" | All | | |
| 1 | Drive extension, $^3/_8$" X IO" | All | | |
| 1 | Drive adapter, $^3/_8$" to '$^1/_4$" | All | | |
| 1 | Drive swivel, $^3/8$" | All | | |
| 1 | Drive extension, $1/_4$" X 6" | All | It is most economical to buy something similar to a 190-piece mechanic's tool set - see last line | |
| 1 | Drive extension, $^3/_8$" X 3" | All | | |
| 1 | Drive extension, 1/2" X 10" | All | | |
| 1 | Drive extension, $^1/_4$" X 3" | All | | |
| 1 | File, triangular, 8" X 1/4" | All | | |
| 1 | File, metal, flat | All | | |
| 1 | File, triangular, 7" X 1/4" | All | | |
| 2 | Hacksaw, carbide blade | All | | |
| 2 | Hammer, carpenter, 16 oz. | All | | |

# Technical Rescue Equipment

Discipline:
1 - Rope/High Angle Operations
2 - Trench Operations
3 - Structural Collapse Operations
4 - Confined Space Operations
5 - Industrial/Agricultural Ops

## ADVANCED HEAVY RESCUE TEAM -TOOL BOX INVENTORY

| Quantity | Item | Discipline | Unit Price ($) | Cost ($) |
|---|---|---|---|---|
| 2 | Hammer, brass, 2 1/2 lb. | All | | |
| 1 | Pipe Wrench, 12" | All | | |
| 1 | Pliers, Channel Lock, 10" | All | | |
| 1 | Pliers, Channel Lock, 14" | All | | |
| 1 | Pliers, 6" | All | | |
| 1 | Pliers, Needle nose, 6" | All | | |
| 1 | Pliers, Electrical, 7" | All | | |
| 1 | Pliers, 8" | All | | |
| 1 | Pry bar, 15" | All | | |
| 1 | Punch, $3/8$" | All | | |
| 1 | Screwdriver set, Phillips head tip<br>1 - 3"  1 - 7"  1 - 8"<br>1 - 10"  1 - 13"  1 - 15" | All | | |
| 1 | Screwdriver set, straight tip<br>1 - 3"  1 - 7"  1 - 8"<br>1 - 10"  1 - 13"  1 - 15" | All | It is most economical to buy something similar to a 190-piece mechanic's tool set - see last line | |
| 1 | Shears, metal, straight cut, 11" | All | | |
| 1 | Socket drive, 1/2 X 10" | All | | |
| 1 | Socket drive, $3/8$" X 10" w/ swivel head | All | | |
| 1 | Socket set, $3/8$", metric sizes 11 mm -19mm | All | | |
| 1 | Socket set, $3/8$", US standard sizes 3/8" - $9/16$" | All | | |

**Technical Rescue Equipment**

Discipline:
1 - Rope/High Angie Operations
2 - Trench Operations

3 - Structural Collapse Operations
4 - Confined Space Operations
5 - Industrial/Agricultural Ops

## ADVANCED HEAVY RESCUE TEAM - TOOL BOX INVENTORY

| Quantity | Item | Discipline | Unit Price ($) | Cost ($) |
|---|---|---|---|---|
| 1 | Socket set, $1/4$", metric sizes 5mm -13mm | All | | |
| 1 | Socket set, $1/4$", US standard sizes $5/32$" - $1/2$" | All | | |
| 1 | Socket set, 1/2", US standard sizes $7/16$" - 1 1/16" | All | | |
| 1 | Socket set, $1/2$", metric sizes 11mm -19mm | All | | |
| 1 | Socket drive, '1/4" X 5" | All | | |
| 1 | Socket drive, 3/8" X 8" | All | | |
| 4 | Spark plugs, spare | All | | |
| 1 | Stapler, T-5O w/ $3/8$" and $5/16$" staples | All | | |
| 2 | Tape, roll, electrical, $3/4$" | All | | |
| 2 | Tape, roll, masking, I" | All | | |
| 2 | Tape, duct, 2" | All | | |
| 1 | Tape Measure, 12 ft. X $3/4$" | All | | |
| 2 | Tape, roll, Teflon, $1/4$" | All | | |
| 1 | Tape Measure, 30 ft. X 1" | All | | |
| 1 | Tool case w/ drawers | All | | |
| 1 | Vice grips, 9" | All | | |
| 2 | Wedges, wood, small | All | | |
| 1 | Wire Cutter, 6" | All | | |
| 1 | Wrench set, combination, metric 6MM – 19MM | All | | |
| 1 | Wrench set, combination, US standard $1/4$" – $1 1/4$" | All | | |
| | Approx. price of 190-piece  tool/chest set | | | 1 ,1 04.00 |

# Technical Rescue Equipment

Discipline:
1 - Rope/High Angle Operations
2- Trench Operations

3 - Structural Collapse Operations
4- Confined Space Operations
5 - Industrial/Agricultural Ops

## ADVANCED HEAVY RESCUE TEAM - TRENCH/STRUCTURAL SHORING UNIT INVENTORY

| Quantity | Item | Discipline | Unit Price ($) | cost ($) |
|---|---|---|---|---|
| 12 | 1" Shoring panels, 41x8" w/ 2"XIZ"XIZ ft. uprights bolted in place | 2,3 | 165 | 1,980 |
| 4 | 2"xl2"x12 ft. uprights, PT | 293 | 13 | 52 |
| 4 | Z'XIZ"X8 ft. uprights, PT | 2,3 | 9 | 36 |
| 20 | Z"X4"XIZ" scabs | 2,3 | 1 | 20 |
| 6 | ³/₁¹ Ground pads, 4x8 | 293 | 42 | 252 |
| 2 | 4"X4"X8 ft. timbers, PT | 2,3 | 11 | 22 |
| 4 | 6"X6"X8 ft. timbers, PT | 2,3 | 28 | 112 |
| 6 | Apron, capenter's | 2,3 | 40 | 240 |
| 3 | Banner guard tape, "Fireline" | 293 | 17 | 51 |
| 2 | Bar, wecking, 30" | 2,3 | 16 | 32 |
| 2 | Bar, pinch, 60" | 2,3 | 21 | 42 |
| 2 | Bar, wrecking, 18" | 2,3 | 9 | 18 |
| 1 | Broom, street | 2,3 | 13 | 13 |
| 6 | Bucket, plastic, 5 gal. | 2,3 | 4 | 24 |
| 2 | Cart w/ CO₂ cylinder, regulator and hose | 2,3 | 175 | 350 |
| 6 | Crate, plastic | 293 | 4 | 24 |
| 1 | Funnel | 2,3 | 6 | 6 |
| 2 | Gas can, 5 gal. | 2,3 | 31 | 62 |
| 8 | Glasses, safety | 2,3 | 6 | 48 |
| 2 | Hammer, sledge, 8 lb. | 293 | 24 | 48 |

**Technical Rescue Equipment**

Discipline:
3 - Rope/High Angle Operations
2- Trench Operations

3 - Structural Collapse Operations
4 - Confined Space Operations
5 - Industrial/Agricultural Ops

| | ADVANCED HEAVY RESCUE TEAM - TRENCH/STRUCTURAL SHORING UNIT INVENTORY | | | |
|---|---|---|---|---|
| Quantity | Item | Discipline | Unit Price ($) | Cost ($) |
| 2 | Hammer, sledge, IO lb. | 2, 3 | 27 | 54 |
| 2 | Hammer, sledge, 4 lb., short handle | 2, 3 | '2 | 24 |
| 4 | Hammer, claw, 22 oz. | 2, 3 | 25 | 100 |
| 2 | Hammer, engineer's, short handle | 2, 3 | 29 | 58 |
| 8 | Hard hat, safety | 2, 3 | 11 | 88 |
| 2 | Hoe, long handle | 2, 3 | 16 | 32 |
| 8 | Knee pads | 2, 3 | 9 | 72 |
| 2 | Ladder, straight, 16 ft. | 2, 3 | 315 | 630 |
| 1 | Nails, box, 16P double-headed | 2, 3 | 50 | 50 |
| 1 | Particle masks, box | 2, 3 | 9 | 9 |
| 9 | Pneumatic shores, telescoping, 18" - 27" | 2, 3 | 250 | 2,250 |
| 9 | Pneumatic shores, telescoping, 23" - 37" | 2, 3 | 289 | 2,601 |
| 9 | Pneumatic shores, telescoping, 30" - 48" | 2, 3 | 329 | 2,961 |
| 9 | Pneumatic shores, telescoping, 42" - 66" | 2, 3 | 365 | 3,285 |
| 9 | Pneumatic shores, telescoping, 60" - 96" | 2, 3 | 400 | 3,600 |
| 9 | Pneumatic shores, telescoping, 90" - 120" | 2, 3 | 445 | 4,005 |
| 9 | Pneumatic shores, telescoping, 114" 144" | 2, 3 | 480 | 4,320 |
| 1 | Post hole digger | 2, 3 | 40 | 40 |
| 1 | Pump, electric, submersible w/ 50 ft. discharge hose | 2, 3 | 360 | 360 |
| 1 | Pump, trash, gasoline-driven, w/two 3" suction and one 3" discharge sleeves | 2, 3 | 940 | 940 |
| 8 | Rope, nylon, 25 ft. | 2, 3 | 6 | 48 |

**Technical Rescue Equipment**

Discipline:
1 - Rope/High Angle Operations
2 - Trench Operations
3 - Structural Collapse Operations
4 - Confined Space Operations
5 - Industrial/Agricultural Ops

## ADVANCED HEAVY RESCUE TEAM -TRENCH/STRUCTURAL SHORING UNIT INVENTORY

| Quantity | Item | Discipline | Unit Price ($) | cost  G) |
|---|---|---|---|---|
| 2 | Rope w/wire slings | 2, 3 | 16 | 32 |
| 2 | Ruler, folding, 6 ft. | 2, 3 | 17 | 34 |
| 2 | Saw, crosscut, hand | 2, 3 | 31 | 62 |
| 24 | Shore swivel ends | 2, 3 | 18 | 432 |
| 3 | Shovel, round point, long handle | 2, 3 | 37 | 111 |
| 1 | Shovel, round point, D handle, short | 2, 3 | 32 | 32 |
| 4 | Shovel, folding, entrenching | 2, 3 | 18 | 72 |
| 1 | Tape measure, 100 ft. | 2, 3 | 38 | 38 |
| 4 | Tape measure, l2 ft. | 2, 3 | II | 44 |
| 3 | Tape measure, 25 ft. | 293 | 14 | 42 |
| I | Tape measure, 50 ft. | 2, 3 | 24 | 24 |
| 2 | Tarp, vinyl, 12X16 | 2, 3 | 17 | 34 |
| 1 | Trl square | 293 | 10 | 10 |
| I 2 | Wedges, mod, 4"X24" | 2, 3 | 2 | 24 |

**Technical Rescue Equipment**

Discipline:
1 - Flatwater Operations
2 - Swiftwater Operations
3 - Ice Operations

## FIRST RESPONDER/OPERATIONS LEVEL WATER RESCUE TEAM

| Quantity | Item | Discipline | Unit Price ($) | Cost ($) |
|---|---|---|---|---|
| 1 | Binoculars | All | 75 | 75 |
| 2 | Blinds for firehose | All | 25 | 50 |
| 4 | Carabiners | All | 18 | 72 |
| 1 | EMS first aid supplies | All | 250 | 250 |
| 1 | Firehose, 50 foot section | All | 120 | 120 |
| 4 | Flashlights, waterproof | All | 25 | 100 |
| 4 | Helmets | All | 26 | 104 |
| 4 | Knives or shears | All | 15 | 60 |
| 4 | Personal floatation devices | All | 50 | 200 |
| 1 | Pike pole or hook | All | 75 | 75 |
| 2 | Rope, life safety/rescue rope, 100 foot (kemmantle) | All | 100 | 200 |
| 2 | Rope, throw bags tith 75 foot rope (polypropylene kemmantle) | All | 75 | 100 |
| 1 | SCBA cylinder, spare | All | 275 | 275 |
| 4 | Webbing, 2, 20 foot | All | 3 | 12 |
| 4 | Whistle | All | 2 | 8 |

**Technical Rescue Equipment**

## ADVANCED LEVEL WATER RESCUE TEAM

| Quantity | Item | Discipline | Unit Price ($) | Cost ($) |
|---|---|---|---|---|
| 2 | Anchor straps | All | 50 | 100 |
| 1 | Backboard and spinal immobilization device | All | 250 | 250 |
| 1 | Barrier tape, roll, "fireline type" | All | 17 | 17 |
| 2 | Binoculars | All | 75 | 150 |
| 4 | Blind caps for firehose, with air connection | All | 50 | 200 |
| 4 | Body bags | All | 35 | 140 |
| 1 | Boogie board | 2,3 | 100 | 100 |
| 4 | Carabiners | All | 18 | 72 |
| 2 | Crampons or cleats, pair | 3 | 75 | 150 |
| 2 | Drag hooks | All | 50 | 100 |
| 2 | Dry suit | All | 400 | 800 |
| 1 | EMS first aid supplies | All | 250 | 250 |
| 4 | Exposure suit | All | 250 | 1000 |
| 4 | Filter masks for body recovery | All | 50 | 200 |
| 1 | Fire extinguisher (per boat) | All | 25 | 25 |
| 1 | Firehose, 50 foot section | All | 120 | 120 |
| 1 | Flare gun | All | 25 | 25 |
| 4 | Flashlights, waterproof | All | 25 | 100 |
| 2 | Floodlights, portable | All | 75 | 150 |
| 1 | Generator, portable | All | 750 | 750 |
| 1 | Global positioning system, handheld | All | 600 | 600 |

**Technical Rescue Equipment**

## ADVANCED LEVEL WATER RESCUE TEAM

| Quantity | Item | Discipline | Unit Price ($) | Cost ($) |
|---|---|---|---|---|
| 4 | Gloves, pair | All | 25 | 100 |
| 4 | Goggles/eye protection | All | 15 | 60 |
| 4 | Helmets | All | 26 | 104 |
| 1 | Hypothermia/exposure kit | All | 200 | 200 |
| 4 | Ice rescue picks | 3 | 10 | 40 |
| 1 | Ice rescue sled | 3 | 700 | 700 |
| 2 | Insect repellent | All | 5 | 10 |
| 4 | Knives or shears | All | 15 | 60 |
| 2 | Lifejacket with "live-bait" quick release | 2 | 125 | 250 |
| 24 | Light sticks, cyalume type | All | 2 | 48 |
| 1 | Line gun | All | 300 | 300 |
| 1 | Litter/stokes basket, with floatation | All | 400 | 400 |
| 1 | Megaphone | All | 100 | 100 |
| 1 | Night vision equipment, infrared | All | 1500 | 1500 |
| 1 | Oxygen and suction unit, portable | All | 510 | 510 |
| 1 | Oxygen heater and humidifier kit | All | 1200 | 1200 |
| 2 | Paddles | 1,2 | 50 | 100 |
| 4 | Personal floatation devices | All | 50 | 200 |
| 1 | Pike pole or hook | All | 75 | 75 |
| 1 | Radio, weatherband | All | 50 | 50 |
| 1 | Reeves stretcher | All | 200 | 200 |

# Technical Rescue Equipment

## ADVANCED LEVEL WATER RESCUE TEAM

| Quantity | Item | Discipline | Unit Price ($) | Cost ($> |
|---|---|---|---|---|
| 2 | Ring buoy | All | 50 | 100 |
| 2 | Rope, life safety/rescue rope, 100 foot (kemmantle) | All | 100 | 200 |
| 4 | Rope, throw bags with 75 foot rope (polypropylene kemmantle) | All | 75 | 300 |
| 2 | Rope pulley | 2,3 | 50 | 100 |
| 2 | Rope rescue harness | 2,3 | 150 | 300 |
| 2 | Salvage tarp, waterproof | All | 100 | 200 |
| 1 | SCBA cylinder, spare | All | 275 | 275 |
| 1 | Shovel, dirt | All | 35 | 35 |
| 2 | Swim fins, mask and snorkel | 2 | 100 | 200 |
| 1 | Triage tags, set | All | 35 | 35 |
| 1 | Underwater sonar system | 2 | 1500 | 1500 |
| 4 | Webbing, Z", 20 foot | All | 3 | 12 |
| 2 | Wet suit | 12 | 200 | 400 |
| 4 | Whistle | All | 2 | 8 |

# BIBILIOGRAPHY

Bechdel, Less, and Slim Ray, *River Rescue*. Appalachian Mountain Club, 1985.

Bentivoglio, John T., "OSHA Compliance." *Firehouse*, May 1995, p. 58-61.

Brennan, Kenneth J., "A Systems Approach to Rope Rescue Safety." *Fire Engineering*, May 1994, p. 49-56.

Collins, Larry, "An Innovative Approach to Swiftwater Rescue." *Firehouse*, May 1994. p. 83-111.

Collins, Larry, "Managing Swiftwater Rescue." *Fire Chief* November 1993, p. 30-38.

Downey, Ray, The *Rescue Company*. Fire Engineering Books and Videos, 1992.

Downey, Ray, "Specialized Rescue Training." *Fire Engineering*, January 1995, p. 24-34.

Gallagher, Tim, and Steve Stormen, "Confined Space Rescue." *Rescue*, May/June 1994, p, 51-69.

Linton, Steven J., and Damon A. Rust, Ice *Rescue*. International Association of Dive Rescue Specialists, 1982.

Malek, Richard S., "The Challenge of OSHA Standard on Permit-required Confined Space Rescue." *Fire Engineering*, September 1994, p. 35-44.

Mellot, Kevin D., "Building a Specialized Rescue Capability on a Budget." *Fire Chief* November 1992, p. 36-39.

Segerstrom, Jim, and Barry Edwards, *Low to High Angle Rescue Techniques*. Rescue 3 International, 1994.

Williams, Rick, "So You Wanna Start a Dive Team." *Rescue*, July/August 1994, p. 32-36.

www.ingramcontent.com/pod-product-compliance
Lightning Source LLC
Chambersburg PA
CBHW081110170526
45165CB00008B/2393